THE PRACTICAL STATISTICIAN

Simplified Handbook of Statistics

Roger E. Kirk, Consulting Editor

EXPLORING STATISTICS: AN INTRODUCTION FOR PSYCHOLOGY AND EDUCATION
Sarah M. Dinham
The University of Arizona

METHODS IN THE STUDY OF HUMAN BEHAVIOR
Vernon Ellingstad and Norman W. Heimstra
The University of South Dakota

AN INTRODUCTION TO STATISTICAL METHODS IN THE BEHAVIORAL SCIENCES
Freeman F. Elzey
San Francisco State University

EXPERIMENTAL DESIGN: PROCEDURES FOR THE BEHAVIORAL SCIENCES
Roger E. Kirk
Baylor University

INTRODUCTORY STATISTICS
Roger E. Kirk
Baylor University

STATISTICAL ISSUES: A READER FOR THE BEHAVIORAL SCIENCES
Roger E. Kirk
Baylor University

THE PRACTICAL STATISTICIAN: SIMPLIFIED HANDBOOK OF STATISTICS
Marigold Linton
The University of Utah
Philip S. Gallo, Jr.
San Diego State University

NONPARAMETRIC AND DISTRIBUTION-FREE METHODS FOR THE SOCIAL SCIENCES
Leonard A. Marascuilo
University of California, Berkeley
Maryellen McSweeney
Michigan State University

DOING PSYCHOLOGY EXPERIMENTS
David W. Martin
New Mexico State University

BASIC STATISTICS: TALES OF DISTRIBUTIONS
Chris Spatz
Hendrix College
James O. Johnston
University of Arkansas at Monticello

MULTIVARIATE ANALYSIS WITH APPLICATIONS IN EDUCATION AND PSYCHOLOGY
Neil H. Timm
University of Pittsburgh

THE PRACTICAL STATISTICIAN

Simplified Handbook of Statistics

MARIGOLD LINTON
The University of Utah

PHILIP S. GALLO, JR.
San Diego State University

With the Assistance of

CHERYL A. LOGAN
University of California, San Diego

BROOKS/COLE PUBLISHING COMPANY
MONTEREY, CALIFORNIA
A Division of Wadsworth Publishing Company, Inc.

To our students, whose problems with research
persuaded us that this book should be written.

ISBN: 0-8185-0127-8
L.C. Catalog Card No.:   73-91423
Printed in the United States of America
15 14 13 12 11 10 9

Production Editor: Lyle York
Interior & Cover Design: Linda Marcetti
Illustrations: Creative Repro Photolithographers, Monterey,
    California
Typesetting: Creative Repro Photolithographers, Monterey,
    California
Printing & Binding: Malloy Lithographing, Inc., Ann Arbor,
    Michigan

PREFACE

When two practical statisticians decide to write a book on statistics, the outcome is not simply a statistics book. Here is a book *about* statistics—a book whose purpose is to help people use statistics as a research tool.

The book was written to satisfy our need for a statistics book to use in our laboratory classes. Most of our students understand a good deal about statistics but lack the skills needed to apply the statistics to actual research situations. Therefore their need is less for a theoretical introduction than for a simple pragmatic approach to recognizing and performing the variety of statistics techniques that they encounter in a laboratory class. We wanted a book that would tell students how to determine *for themselves* what analysis was appropriate and then detail how to perform the computations for the particular designs. The students' increased competence, confidence, and self-reliance was an intended and exciting side-effect of such an approach.

We believe we have written a book that satisfies the needs of the struggling, inexperienced researcher-statistician. It is written for the student who, like ours, has had at least some minimum exposure to—but has not necessarily mastered—elementary statistics. The book may be particularly helpful (1) as the statistical guide for undergraduate or graduate research-methods classes and (2) as a supplemental book in a basic or advanced statistics course that emphasizes the practical or applied aspects of inferential statistics.

Some of the features of the book can best be understood in terms of a behavioral-objectives framework. Our ultimate objective is for students *to analyze their data correctly, quickly, and comprehensively and to provide proper statistical and experimental interpretation of their outcome.* The intermediate objectives are most simply seen in the implicit structure of the book and in the explicit structure of the Branching Program. The student must be able to identify the independent and dependent variables in a given study, must be able to determine the appropriate statistical procedure, must know the kind of data collected, and so on. Aside from basic computational skills, statistical analysis involves a variety of complex decisional skills. We have attempted to focus readers' attention on the decisions that must be made when an analysis is to be performed. Since we assume that students need explicit training in the individual elements involved, we break the decision-making process into a series of steps. The student then has a simple set of easily learned routines to perform. Let's return to our major behavioral objectives for students:

1. To analyze the data *correctly*. Our Branching Program feature makes explicit, and breaks down into discrete units, the steps involved in making decisions about the statistic to be employed. With some practice every reader can learn to arrive at the correct analysis every time.

2. To analyze the data *quickly*. Every analysis is presented so that the student who has mastered the computational procedures may follow them easily. The student who has not performed the computations previously or who has difficulty in performing computations is provided with a simple verbal statement of what the computations are and with step-by-step procedures, usually in the form of a worked example.

3. To analyze the data *comprehensively*. Every analysis is followed by explicit reference to specific comparisons, if these are necessary, and to strength-of-association measures, reminding the reader that obtaining an answer to the test of significance does not necessarily complete the task of analyzing the data.

4. Finally, to provide proper statistical and experimental *interpretations*. Each analysis explicitly directs the reader to the important chapter on interpretation. In addition, each example is accompanied by a statistical and an experimental interpretation to provide additional models.

Although some sufficiently motivated undergraduates will be able to proceed through this book with little outside help, we have found that almost all students can become competent at making statistical decisions if they are given repeated practice in making the discriminations required. An effective technique is to provide a series of descriptions of real-life experimental problems, graded from easy to difficult, and accompanied by a series of questions, also graded in difficulty, that focus on each of the decisions that the student must make in following the Branching Program or in making interpretations.

We wish to emphasize that this book cannot substitute for (although it may augment) a thorough background either in experimental design or in advanced statistics. A grounding in both of these topics is vital for a well-rounded research scientist. We are convinced, however, that students can be helped to use statistics appropriately before they understand completely the theoretical underpinnings of such techniques. Furthermore, if students can perform statistical analyses without undue agony, they will be more likely to appreciate statistics as a friendly tool and be encouraged to learn its theoretical bases and the subtleties of their application. Because our emphasis is on usefulness, this book contains no derivations, no formulas that are not directly used in the computations, and only a small amount of theoretical and introductory material. The basic statistical terms used in this book are defined in Chapter 2, *Terms and Definitions*. Those definitions that are essential to using the Branching Programs are marked with an asterisk.

We wish to express our appreciation to the following institutions for their direct or indirect assistance in this project: to San Diego State University, which provided a sabbatical leave to each of us the year that most of this book was written, and to the Center for Human Information Processing, University of California, San Diego, which provided Marigold Linton with a stimulating and congenial home during this year.

We would like to express special appreciation to Norman H. Anderson, whose elegant theoretical approach was instrumental in shaping our own conceptions of statistics. As a theoretician, he is not embarrassed by our public acknowledgement that he taught us "all we know"; but, as practical statisticians, we are the first to protest the converse implication, that he taught us "all he knows." We would also like to thank Professors S. Joyce Brotsky of California State University at Northridge, Daniel L. Lordahl of Florida State University, Charles G. McClintock of the University of California at Santa Barbara, and, especially, Eugene Zechmeister of Loyola University, Chicago, for their helpful reviews of the manuscript. Finally, our special thanks to Brooks/Cole's consulting editor, Roger E. Kirk, for his excellent criticism.

We gratefully acknowledge the help of our colleagues Edward Alf, Jr., Al Hillix, Elizabeth Lynn, and James Ohnesorg, who have assisted us with a number of general and specific problems in the presentation of various statistics. Our careful typist, Teddy Ralph, astounded us with the speed with which she turned out the hundreds of virtually errorless pages of this very complex text. We owe her many thanks.

And our thanks to our students, who, struggling through an earlier draft of the book, have flattered us by understanding more about statistics than we could reasonably expect.

Marigold Linton
Philip S. Gallo, Jr.

CONTENTS

CHAPTER ELEVEN / STRENGTH-OF-ASSOCIATION MEASURES   329

CHAPTER TWELVE / CORRELATION COEFFICIENTS AND SIGNIFICANCE TESTS FOR CORRELATION COEFFICIENTS   339

CHAPTER ONE

INTRODUCTION

Perhaps the most difficult task, when we face new students, is to persuade them that statistics (despite its fearsome reputation) is simpler than it looks and that statistical tests are there to help them, not as a roadblock. This book provides a simple framework that will permit you, once you are familiar with it, to perform a wide variety of simple analyses. Most importantly, it tells you *how to figure out which analysis to perform*. The routines are very simple once you are familiar with them; however, since there is always grave risk that the terrified explorer will miss the forest for the trees, you should read the next section.

HOW TO USE THE BOOK

The book is designed to help you analyze your data correctly, quickly, and comprehensively and to provide proper statistical and experimental interpretations of your outcomes. It does so by providing a decision-making framework from which you determine which analyses are appropriate for your data. It presents step-by-step procedures for each of the most common analyses, including both the computational formulas and (in most cases) a worked-out example. Finally, when you have completed your basic test of significance (the most common choice), the book first directs you to a detailed statement of how to interpret your outcomes and then suggests additional computations that may assist you in providing a complete interpretation of your study.

If you are inexperienced with performing analyses, and, in particular, with deciding which is the appropriate analysis, we would strongly urge that you use the book in the following way.

First, after you have read this introductory chapter, read Chapter 2, *Terms and Definitions*. Pay particular attention to the "branching program" at the end of the chapter, because your next reading material depends on your decision here. If the statistical technique needed is the significance test, read Chapters 3 and 4. If you require a correlation coefficient, read Chapter 12. Chapters 3, 4, and 12 provide general introductory material and should be reread on each usage until you are thoroughly familiar with the information. Statistical information may be quite clear at the time that you read it, but distinctions blur quickly with the passage of time. After a number of readings the information will become "yours," but it is essential that you review the information until it is accurately stored in your memory.

When you have completed these introductory chapters, return to the branching program and then proceed to the specific chapter that describes your statistical test. Each of these specific chapters follows the same pattern: it begins with a branching program that directs you, as you make a series of decisions, to the specific analysis appropriate for your data. Accompanying each branching program are a set of brief *Limitations and Exceptions* that provide cautions and suggest difficulties for the particular tests involved. The *Limitations and Exceptions* should be read thoroughly when you use the program for the first time, but they should be checked at least briefly each time you perform computations. Following the *Limitations and Exceptions*, a somewhat longer section discusses how the particular statistics work, the problems in performing such analyses, and notation. You should read this section thoroughly at least once before you perform the statistics in any chapter; thereafter you will probably rarely need to consult it. Finally, each chapter contains step-by-step computational procedures and (in most cases) a worked example.

The computational procedures are designed for readers with various backgrounds. The most advanced reader may be able to glance at the computational formulas or summary tables and perform the computations. For this reason, the computational formulas are placed conveniently at the beginning of the steps for every analysis. If you require more specific guidance, each analysis is accompanied by step-by-step computational formulas and by a verbal statement of the computations required. For almost all analyses there is also a step-by-step worked example illustrating the general computational formulas.

Following each analysis there are three references: to the chapter on interpretation appropriate to this analysis, to the chapter on specific comparisons, and, finally, to a chapter on strength-of-association measures. In addition, an interpretation of the study follows each worked example.

FEATURES OF THE BOOK

The division of the book into tests of significance and correlation coefficients reflects a major difference in the basic kinds of analyses and, more importantly for our framework approach, in the kinds of variables that must be considered in deciding which kind of analysis is appropriate. Although we include many tests of significance, we include relatively few correlation coefficients. This decision reflects the needs of most researchers rather than the relative development of these areas.

The number of statistical tests we have included may seem formidable to the student. From the viewpoint of the researcher, however, the variety of statistics is rather limited. We have attempted to be reasonably comprehensive in our coverage of simple statistical analyses, but we have generally avoided the "elegance" of providing two analyses that do the same thing. Wherever possible, we have included either the most powerful or the most general techniques appropriate for solving a specific research problem. For instance, although there are many statistical tests for frequency data, we have presented only the χ^2 test because it can be expanded to handle most situations that arise with frequency data. For the same reasons, we recommend only analysis of variance for score data (although we have included the t test). In summary, we have attempted to cover a large number of research designs with a few well-selected tests. Our informal surveys of the research literature in several fields suggest that the techniques presented in this book might have been used appropriately, or were actually used, to analyze the results of 85 to 90% of these published studies. To have included the techniques used in the remaining 10 to 15% of the studies would have more than quadrupled the length of this book. It seems clear that a relatively small number of statistical techniques can handle most research problems. The small number of research designs that cannot be handled by these tests require a wide variety of more specialized statistical tests.

A final point on coverage deserves emphasis. Because we conceive this book primarily as a guide for laboratory research, we do not believe that the student need necessarily master, or even be exposed to, every statistic presented in the book. We recommend that you be aware of the alternatives, be comfortable enough with moderately complex designs that you don't discard them simply out of fear, and that you then *select the simplest design that will answer your research*

question. Again, your most important task is mastering the decisions
made in selecting an analysis. Having learned this framework thoroughly,
you should be able to deal competently with more complex material
you encounter later.

Most students in a laboratory class should expect direct research
experience with not more than half a dozen of the simpler designs.
(It should be noted, though, that for a class of novice researchers the
choice of designs may vary widely.) Although every student should
recognize that the more complicated designs are simply extensions
of simpler, more familiar designs, most of the complex designs are
presented largely for reference and to furnish a clear example of the
scope of such designs. In short, the complex analyses presented in
this book should be used only if they are demanded by your research
question; complexity is not something to be sought for its own sake.

Only a knowledge of simple algebra is required in order to use
this book. We have included no theoretical derivations. To further
simplify the reader's task, all computational steps are accompanied
by simple verbal equivalents. Although we recognize that many
researchers compound their work many times by performing inef-
ficient computations, we have reluctantly omitted specific notes on
the use of calculators or computers. It is impossible in this period of
rapid technological change to guess what kind of computers or
calculators the reader will use. For any of the complex or lengthy
statistical computations, however, major time-saving procedures may
be followed. For example, most machines permit sums and sums of
squares to be accumulated simultaneously, thus saving much time in
both the analysis of variance and correlation. The time taken to consult your calculator or computer guide for helpful computational
hints will be well repaid.

Finally, we would like to point out, chapter by chapter, the
major features of the book. Chapter 2, as we have indicated, presents
brief basic definitions to be used in the book. It also includes the
initial branching program that may be used following Chapter 2 or
Chapter 3, depending on your degree of experience.

Chapter 3 is designed for those who need additional study on
the terms in Chapter 2. It provides specific research examples and
then analyzes them along the dimensions required for making
decisions about specific statistical analyses. In essence, we set out
five dimensions along which research designs may vary. Once you
can appropriately apply these terms to your research design, the
branching program indicates the analysis technique you should use.

Chapter 4 is not vital to the use of the branching program or to
the appropriate analysis of data, but we believe that every student

should review it briefly before attempting computations or interpretations. This chapter explains the workings of statistical tests of significance and also discusses such basic concepts as the null hypothesis, Type I and Type II errors, alpha level, one-tailed vs two-tailed tests of significance, and populations. A thorough reading of this chapter will assure that the reader performs the correct test of significance and properly interprets his findings.

Chapters 5 through 8 present the major statistical tests of significance. The chapters are organized around the kind of data used (the dependent variable). Chapter 5 presents χ^2 tests for frequency data, Chapter 6 presents a variety of tests for ordered data, and Chapter 7 and 8 present the analysis of variance and the t test for score data. Brief introductory material at the beginning of each chapter indicates the nature and problems of the tests presented.

Chapter 9 presents detailed procedures for interpreting the results of each analysis. It provides the reader with a simple, logical method for systematically interpreting results. The orderliness of such interpretational procedures becomes crucial in large or complex studies; for these we have provided procedures for developing a proper interpretation for most outcomes of these studies. Since it is often impossible to interpret correctly the meaning of an interaction in an analysis of variance unless the results have first been graphed, this chapter emphasizes graphical data presentation. Procedures for transforming tabular data into graphical data and examples of graphical interpretation are presented.

Chapter 10 presents statistical tests that are used to make specific comparisons between selected groups or conditions after the overall analysis has been performed. In complex research designs, the overall analysis often answers such a general question about the relationship of the variables that the researcher may not be satisfied that he understands the relationship among his variables. Specific-comparison tests allow the researcher to determine the locus of specific differences in the data. Chapter 10 discusses the special problems associated with the use of specific comparisons and suggests appropriate tests and computational rules.

Chapter 11 presents strength-of-association measures. More and more researchers are realizing that significant relationships are not necessarily large or important. The material in Chapter 11, although far from complete, gives you some basic techniques for assessing the strength of the effect you have uncovered. We would guess that within 10 years journals will require that all statistical tests of significance be accompanied by some descendant of these strength-of-association measures. This area is still in its infancy, and during the next few years there will certainly be a marked increase in the sophistication of these techniques. We present those techniques on which there is already some agreement.

Chapter 12 provides an overview of the various kinds of correlational techniques, followed by the same comprehensive step-by-step coverage described for the earlier chapters. In addition, Chapter 12 presents significance tests for all of the correlational techniques. Just as we recommend that all tests of significance should, if possible, be accompanied by a strength-of-association measure, we also recommend that all correlation coefficients be accompanied by a statement of their significance or lack of significance.

CHAPTER TWO

TERMS AND DEFINITIONS

This chapter provides the reader with thumbnail definitions and examples of a number of basic terms. Everyone who does research should be familiar with these terms, and mastery of these concepts is necessary for the effective use of this book. Readers who are well grounded in basic statistics and experimental design should be able to use the branching programs provided in this and in subsequent chapters immediately after reading this chapter.

If the brief descriptions and examples given in this chapter do not make the concepts clear, the following procedures will be helpful: If a significance test is the statistical technique required, read Chapters 3 and 4 carefully. If a correlation coefficient is the statistical technique required, read the introduction to Chapter 12. These chapters provide more detailed explanations and further examples of the basic concepts.

Terms essential to the use of the branching program are marked with an asterisk.

RESEARCH

Research is the empirical investigation of the relationships between or among several variables. In the behavioral sciences, the relationships of interest are generally those that exist between some behavior of the organism (a response variable) and other variables (stimulus variables, task variables, other response variables, and so on) that may be related to that behavior.

EXPERIMENTAL RESEARCH

In experimental research, a variable of some type is actively manipulated (changed in value) by the researcher in order to determine whether there are concomitant changes in a behavior of interest. In addition to the requirement that a variable be manipulated by the experimenter, a well-controlled experiment requires that the assignment of subjects to conditions, or of conditions to subjects, be made on a random basis and that all other variables except the independent variable or variables (the variable or variables that are manipulated) be held as constant as is possible in the given situation.

NONEXPERIMENTAL RESEARCH[1]

In nonexperimental research, the researcher does not manipulate a variable. All research that involves relationships between or among various response variables falls into the category of nonexperimental research. For example, much research that is conducted in natural settings is nonexperimental, since the researcher is unable to manipulate the conditions that the subjects will experience.

*INDEPENDENT VARIABLE

Strict definition: The independent variable is the variable manipulated by the experimenter. The independent variable is under the experimenter's control and he can determine the values it will assume. If there is a relationship between the independent variable and the behavior being studied, different values of the independent variable should produce differential effects on the behavior being studied. In other words, a group of subjects who have experienced one value of the independent variable should behave differently from a group of subjects who have experienced a different value.

Strictly speaking, this definition applies only to experimental research. However, since the same statistical techniques can often be applied to the results of both experimental and nonexperimental research, a broader definition of "independent variable" is used in this book.

Broad definition: An independent variable is any variable, regardless of type, that is assumed to produce an effect on, or be related to, a behavior of interest.

Examples. If the behavior of interest is performance on a mathematics test, independent variables that might be studied include:
Amount of sleep the night before the test
IQ of the subjects
Room temperature
Number of years of prior mathematics training
Length of time allotted in which to complete the exam

---

[1] Some research may involve both experimental and nonexperimental factors.

*NUMBER OF LEVELS OF AN INDEPENDENT VARIABLE

An independent variable may assume any number of different values at the experimenter's discretion. Each value is known as a "level of the independent variable." *The simplest possible research design consists of one independent variable with two levels. Unless there are at least two levels of an independent variable, no comparison between conditions is possible and therefore no decision may be made concerning the effect of the independent variable on the behavior being studied.*

Examples. A study is conducted to determine the effect of room temperature on test performance.

If the experimenter tests the subjects at 70, 80, and 90°, there is one independent variable (room temperature) with three levels.

If the experimenter tests the subjects at 70, 80, 90, 100, and 110°, there is one independent variable (room temperature) with five levels.

If the experimenter tests the subjects at 70, 80, and 90°, and, in addition, half of the subjects had no sleep the night before and the other half had 8 hours' sleep, there are two independent variables (room temperature, sleep deprivation). Room temperature has three levels; sleep deprivation has two levels.

FACTORIAL DESIGN

A factorial design allows two or more independent variables to be combined in the same study so that the effects of each variable can be evaluated independently of the effects of the other(s). In addition, a factorial design allows the experimenter to determine whether there was an interaction between his independent variables— that is, whether the effects of one variable depend on the specific level of the other variable with which it was combined.[2] All statistical techniques in this book that are designed for a single study with more than one independent variable require a factorial design. To produce a factorial design, each level of one independent variable is paired with each level of the other independent variable(s).

Example. Suppose we have a simple factorial design with two independent variables, room temperature and number of minutes allotted to complete a test. Room temperature has two levels, 70 or 90°. Number

---

[2]The concept of interaction is often very difficult for the beginning researcher to grasp. It is treated in detail in Chapter 9.

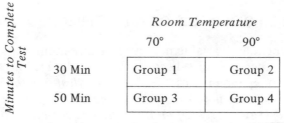

Room Temperature

		70°	90°
30 Min		Group 1	Group 2
50 Min		Group 3	Group 4

Minutes to Complete Test

of minutes to complete the test also has two levels, 30 or 50 minutes. Note that in order to have a factorial design with two levels of each of two independent variables, four research conditions are needed: 70°, 30 min; 70°, 50 min; 90°, 30 min; and 90°, 50 min. The evaluation of the effects of room temperature involves comparing the combined performance of Groups 1 and 3 with the combined performance of Groups 2 and 4. If the number of subjects in each group is equal, this comparison is unaffected by the time-pressure variable, since it is balanced out. Similarly, the effects of time pressure are evaluated by comparing the combined performances of Groups 1 and 2 with the combined performances of Groups 3 and 4. This comparison is unaffected by the room-temperature variable.

DEPENDENT VARIABLE

The dependent variable is the measure of the behavior that the researcher observes but does not manipulate or control. Changes in the dependent variable are brought about by the different values of the independent variable(s) used in the study. Although several dependent variables may be measured in a single experiment, all statistical techniques presented in this book may be applied to *only one dependent variable at a time*.

Examples. If the researcher is studying the effects of different modes of presentation of paired associates (auditory, visual) on learning, dependent variables that he might wish to study include:
Number of trials to some criterion of correct response
Latency of response
Number of errors to some criterion
Amount of time required to learn

*TYPE OF DEPENDENT VARIABLE

In all research, the dependent variable is quantified in some fashion. The statistical analysis is carried out on the numerical values of the dependent variable. In most cases, the quantification process yields one of three basic types of data: score data, ordered data, or frequency data.

Score Data

Each subject is assigned a numerical score that represents his performance or behavior. Score data generally require relatively precise measuring instruments and some degree of knowledge of the behavior being measured.

Many statistical techniques that have been developed to analyze score data make rather stringent assumptions about the nature of the scores. The most common assumptions are:

1. The intervals between scores are equal; that is, differences between scores at one point of the measuring scale are equivalent to the same size differences at any other point on the scale.
2. The scores are assumed to be normally distributed within the population or populations from which they were drawn.
3. The variances of the populations are assumed to be homogeneous.

However, empirical research, particularly with the analysis of variance, indicates that violations of these assumptions do not usually impair critically the usefulness of these tests.

Ordered Data

Each subject is assigned a rank that represents his position along some ordered dimension. Ordered data are obtained when reliable score data cannot be (or are not) obtained, but the subjects can be ranked from high to low along the dimension of interest. In some cases, a researcher may convert score data to ranks because he believes that his measuring instrument was not precise enough to enable him to trust the numerical scores or because he believes that the assumptions underlying a statistical test for score data would be badly violated by his data. Statistical tests designed for use with ordered data generally do not make stringent assumptions about the nature of the underlying distributions and hence are more conservative than those designed for score data.

Frequency Data

Each subject is counted as being in a particular classification, or cross-classification, of categories. For the correlation coefficient, either the frequency of occurrence of subjects in categories or the categories themselves may provide the data. For tests of significance, however, the *frequency of occurrence* of subjects in each category always provides the data. Such data are particularly common in the many types of research in which assigning scores or ranks to subjects would be meaningless.

Example. If the researcher were interested in determining whether there was a relationship between religious preference and political party affiliation, he might establish certain categories of religious preference (for example, Catholic, Protestant, Jewish) and certain categories of political party affiliation (for example, Democratic, Republican, American Independent). He would then note the frequencies of people in each cross-classification to see if some categories were over-represented and others under-represented. It should be emphasized that to assign *a score* to an individual's political party affiliation or to his religious preference would be quite arbitrary and that the results of any statistical test based on such assignments would probably be meaningless.

*BETWEEN-SUBJECTS DESIGN (COMPLETELY RANDOMIZED)

In a between-subjects design, all comparisons between different conditions are based on comparisons between *different* subjects. Each subject serves in only one research condition and contributes only one score to the analysis. That is, each subject experiences only one level of the independent variable, or, in a factorial experiment, one combination of levels.

Example. Twenty subjects took a mathematics aptitude test with the room temperature set at $70°$, and a *different* group of 20 subjects took the test with the room temperature at $90°$.

*WITHIN-SUBJECTS DESIGN (REPEATED MEASURES)

In a within-subjects design, all comparisons between different conditions are based on comparisons within the *same* group of subjects. Each subject serves in all experimental conditions and contributes as many scores to the analysis as there are experimental conditions. That is, each subject experiences all levels of the independent variable or, in a factorial experiment, all combinations of levels.

Example. Twenty subjects took a mathematics aptitude test after 2 hours' sleep and the *same* 20 subjects took a comparable test after 8 hours' sleep.

Matched-Groups Designs

There is an additional class of designs, matched-groups designs, that lie between between-subjects designs and within-subjects designs. Each of these designs shares a basic strategy: to have groups that are essentially equal to begin with so that the effect of an imposed condition may be evaluated. The between-subjects design accomplishes this through randomization; the within-subjects design, by having the same subjects serve in all conditions. The matched-groups designs obtain equality by assigning subjects matched on some dimension to the various conditions. If the factor on which the subjects are matched is strongly correlated with the dependent variable, this may be a very effective design. Examples of matched-groups procedures are the use of litter-mates in animal research and the use of identical twins in human research. The computations are not difficult, but they have been omitted from this discussion because they are used relatively infrequently. Computational procedures may be found in Edwards (1968, p. 155 ff).

*MIXED DESIGN[3]

A mixed design combines a between-subjects design and a within-subjects design. All mixed designs must have at least two independent variables, one that involves a between-subjects comparison and another that involves a within-subjects comparison. Thus some comparisons in a mixed design involve *different* groups of subjects and other comparisons involve the *same* group of subjects. Only mixed designs that involve a factorial combination of independent variables are covered in this book.

Example. Twenty subjects took a mathematics aptitude test with the room temperature set at 70° after 2 hours' sleep and took a comparable test at the same room temperature after 8 hours' sleep. A different group of 20 subjects took the test with the room temperature set at 90° after 2 hours' sleep and took a comparable test at the same room temperature after 8 hours' sleep.

It may be seen (below) that the comparison between 70° and 90° involves a comparison of the performance of Group 1 with Group

---

[3] Although many statisticians use the term "mixed design" in the same way we do, some use it to denote a research design that includes the factorial combination of at least one independent variable that is a fixed factor and one that is a random factor. See Chapter 7 for the definition of fixed and random factors.

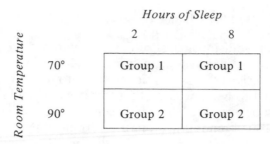

2, a between-subjects comparison. The comparison of 2 hours' sleep and 8 hours' sleep involves a comparison of the performance of Groups 1 and 2 after 2 hours' sleep with the performance of Groups 1 and 2 after 8 hours' sleep, a within-subjects comparison.

POPULATION

A population is the total universe of all possible observations that can be identified by a given set of rules. Populations can be finite or infinite, theoretical or real. Descriptive statistics, such as the mean, median, standard deviation, and so on, are known as *parameters* when they are based on all of the values in a population. Population parameters are seldom actually measured, but they may be estimated from the values of a sample drawn from that population.

Examples:
The IQ scores of all adult females living in the United States
The outcomes of an infinite series of flips of a fair coin
The grade-point averages of all freshmen at Midwest University

SAMPLE

A sample is a subset of observations drawn from a given population.

Examples:
The IQ scores of 104 adult females living in the United States
The outcome of 20 flips of a fair coin
The grade-point averages of 18 freshmen at Midwest University

RANDOM SAMPLE

A random sample is a subset of observations drawn from a given population in such a way that each observation contained in the population has an equal chance of being included in the sample. In practice, samples seldom meet this criterion for randomness, but they are treated as random if no systematic bias exists that might be expected to invalidate the generalizations based on the sample.

RESEARCH HYPOTHESIS

A research hypothesis makes a statement concerning a presumed relationship between an independent variable or variables and a dependent variable. The research hypothesis may be derived from a theory, may be based on past observations, or may be merely a "hunch." It is an expression of the researcher's expectations concerning the outcome of his research and is instrumental in guiding his selection of independent and dependent variables. The research hypothesis cannot be tested directly and must be translated into a directly testable statistical hypothesis.

Examples:
"The rats that were food-deprived for 24 hours will make fewer errors in the maze than the rats that were food-deprived for only 4 hours."

"The children who experience democratic leadership will have higher morale than the children who experience autocratic leadership."

"The average annual income of university graduates will be higher than the average annual income of high school graduates."

(QUESTION: Can you identify the independent and dependent variables for each of these hypotheses?)

STATISTICAL HYPOTHESIS

The statistical hypothesis makes statements about the value of certain population parameters. Statistical hypotheses are directly related to (logically derived from) the research hypothesis, so that acceptance or rejection of the statistical hypotheses allows the researcher to evaluate the truth or falsity of his research hypothesis. The statistical hypotheses evaluated in research are the null (H_0) and the alternative (H_1) hypotheses.

NULL HYPOTHESIS (H_0)

The null hypothesis makes a statement about the value of certain population parameters and is generally phrased to negate the possibility of a relationship between the independent and dependent variables. If the null hypothesis is true, the possibility that the research hypothesis is also true is negated. The null hypothesis is assumed to be true unless the results of a statistical test of significance permit it to be rejected.

ALTERNATIVE HYPOTHESIS (H_1)

The alternative hypothesis makes a statement about the value of certain population parameters and is phrased to contradict the null hypothesis. H_0 and H_1 are mutually exclusive and exhaustive. Rejection of the null hypothesis leads to acceptance of the alternative hypothesis and to the possibility that the research hypothesis is true.

NOTE: Beginning researchers should not allow the frightening language of statistical hypotheses to blind them to their rather simple meaning. Virtually all research is conducted on samples, whereas generalizations from the research are applied to populations. Researchers are not interested in concluding merely that the group of rats deprived of food for 24 hours learned the maze faster than the group of rats deprived of food for 4 hours; they are interested in generalizing their findings to the population of rats from which the sample is drawn. In other words, they wish to conclude that this population of rats learn the maze faster when they are 24-hours deprived than when they are 4-hours deprived. They therefore make some assumptions about population values and determine whether the results of their samples lend credence to these assumptions.

In the example above, the research hypothesis would probably be, "The 24-hour-deprived rats learn the maze faster than the 4-hour-deprived rats." The researchers then randomly assign rats to either a 24-hour food-deprivation condition or a 4-hour food-deprivation condition. In effect, they have drawn a random sample of rats from a population of 24-hour-deprived animals and a second random sample from a population of 4-hour-deprived animals. Their most probable null hypothesis is "$\mu_1 = \mu_2$" ($H_0 : \mu_1 = \mu_2$). This statement means that the mean of Population 1 (of all possible scores of rats deprived 24 hours) equals the mean of Population 2 (of all possible scores of rats deprived 4 hours). If the null hypothesis were true, we would expect little difference in performance between the two groups. A probable alternative hypothesis would be "$\mu_1 \neq \mu_2$" ($H_1 : \mu_1 \neq \mu_2$). In other words, the means of the two populations are not equal and we would expect this inequality to be reflected in the means of the two groups of rats. If the statistical test of significance indicates that there is a very low probability that the sample values would occur if H_0 were true, then H_0 would be rejected and H_1 would be accepted. The research hypothesis would be considered supported if, and only if, the

mean scores of the group deprived 24 hours in fact reflected superior performance as compared to the group deprived 4 hours. The null hypothesis
would also be rejected if the 4-hour group performed better than the 24-
hour group. Such an outcome would not lend support to the research
hypothesis.

*STATISTICAL TEST OF SIGNIFICANCE (SIGNIFICANCE TEST)

A statistical test of significance allows the researcher to evaluate
the probability that the observed sample values would occur if the null
hypothesis were true. If that probability is sufficiently low, the researcher feels justified in rejecting the null hypothesis (and thereby
accepting the alternative hypothesis). More simply stated, a test of
significance provides the researcher with some evidence for concluding,
with a specified risk of error, that there are or are not real differences
between conditions in the population.

TYPE I ERROR

You commit a Type I error when you reject a true null hypothesis.
Usually this involves deciding that a real difference exists in the population when no difference does exist.

ALPHA LEVEL (α)

The probability of rejecting a true null hypothesis is called alpha.
The experimenter controls the probability of making a Type I error
in his study by selecting his own alpha level *before running the study*.

TYPE II ERROR

You commit a Type II error when you fail to reject a false null
hypothesis. Usually this involves deciding that no difference exists
when a real difference does exist.

BETA LEVEL (β)

The probability of accepting a false null hypothesis is called beta.
The beta level for any experiment cannot be precisely controlled;
neither can its size be easily estimated. However, the beta level is
inversely related to (1) the alpha level, (2) the number of subjects in
the experiment, (3) the size of the experimental effect (the strength

of the relationship between the independent and dependent variables), and (4) the variability associated with extraneous variables. The experimenter can reduce the probability of making a Type II error (1) by using an alpha level of .05 rather than a more stringent one, as .025 or .01, (2) by using as many subjects as can be reasonably obtained, (3) by selecting the levels of the independent variable so as to maximize the size of the expected effect, and (4) by reducing extraneous variability (controlling more variables).

*CORRELATION COEFFICIENT

A correlation coefficient expresses the strength of a relationship between two variables. Traditionally, correlation has most often been used to express the strength of a relationship between *two sets of scores* obtained from the same subjects. However, correlation coefficients can be used to assess the strength of a relationship between any two observations that have been made on the same subjects, regardless of whether they are scores, ranks (ordered data), frequencies (categories), or any combination of these.

Since a correlation coefficient denotes the strength of a relationship between two variables in a sample or set of samples, it is both possible and desirable to apply a test of significance to the coefficient. The usual null hypothesis tested is that the correlation in the population from which the sample(s) was drawn is 0. If the null hypothesis is rejected, the researcher can then accept the alternative hypothesis that the variables are correlated in the population.

STRENGTH-OF-ASSOCIATION MEASURE

The strength-of-association measure expresses the strength of a relationship between an independent and a dependent variable. (It should be emphasized that a statistical test of significance does not provide information on strength of the relationship regardless of the alpha level at which the null hypothesis is rejected.) The strength of the relationship is generally reported in the form of a squared correlation coefficient. In many cases, strength-of-association measures can be simply and directly computed from the numerical value of the test of significance (see Chapter 11).

Although it has not yet become common practice, we strongly recommend that all researchers report not only the results of the statistical test of significance but also the value of the strength-of-association measure if the null hypothesis has been rejected. The combination of these two statistical techniques provides the researcher with important information to aid him in evaluating the research hypothesis.

SPECIFIC-COMPARISON TEST

A specific-comparison test allows the comparison of pairs of means, medians, or frequencies within the context of a larger research design. When an independent variable has several levels, rejecting the overall null hypothesis does not always provide precise information concerning which groups are different from which other groups. Specific-comparison tests allow the researcher to pinpoint the group differences.

INTERPRETATION

An interpretation of a study translates statistical findings into statements concerning the relationship between independent and dependent variables within the context of the research problem studied.

WHAT STATISTICAL TECHNIQUE IS NEEDED?

The first question to ask in determining the appropriate analysis to perform on your data is: what statistical technique is needed? You must choose between significance tests and correlation coefficients. In experimental research, you will ask about significance differences more often than you will ask about degree of relationship between variables; therefore you will use significance tests more often than correlation coefficients. Don't guess, though! Be sure that you can distinguish the kind of question you are asking in your research and the kind of statistical technique you will need. During the time you are learning these distinctions, you may make errors. Be alert to the possibility of a mistake on this question; if any later questions in the branching program don't seem reasonable, recheck your choice here.

Suppose you have a group of subjects who are instructed to learn one list of words using rote memory procedures and are then asked to learn a second list of words by creating lively images to relate the words to each other. You think the list that is learned using lively imagery will be remembered better than the one learned by simple rote memory. Are there, in fact, differences between the conditions? You require a statistical test of significance (significance test) to answer this question. Any study that implicitly or explicitly involves the comparison of two or more groups or conditions requires a significance test.

Other examples of questions that require significance tests are: Do dogs learn tricks faster than cats? Do people with high levels of anxiety seek psychotherapy more often than people with low levels of anxiety? Do rats trained under bright lights run faster than those trained under dim illumination? Are people from high socioeconomic groups more likely than those from low socioeconomic groups to prefer Candidate X?

There is a subtle but very real difference between these questions, which may be answered by performing a significance test, and the questions that require a correlation coefficient. Suppose you have a group of subjects for which you have two sets of scores: a set of anxiety measures and a set of scores on a learning task. You wish to know whether anxiety is related to learning. Note that in this case you cannot rephrase the question to ask whether anxiety is different from learning.[4] Anxiety probably is different from learning, but that question is not the same as your original question. Here you wish to know the degree of relationship between these two variables; thus the appropriate statistical technique is the correlation coefficient.

How can this kind of research be distinguished from the former? One critical variable is that you must have paired scores. Usually these paired scores are from the same individual, but sometimes they are from other logical pairs (see Chapter 12). A good check, then, is to see whether there are indeed paired scores. In addition, you should be seeking a relationship between two sets of variables, and your question must not be rephrasable as a question about the difference between groups.

The following studies would be correctly analyzed by a correlation coefficient. (1) You have obtained the weights of ten rats and you know the number of trials each required to learn a maze. What is the relationship between weight and trials required to learn? (2) You have the IQ scores for 20 husband-and-wife pairs. You wish to know whether the IQs of husbands and wives are related. (However, note that, if you wish to determine whether husbands are duller than their wives, you would have to perform a test of significance.)

If the distinction between statistical techniques is clear, you are now ready to start through the branching program. If the distinction

---

[4]Suppose, however, that you are really interested in the question: do persons with high levels of anxiety learn less well than persons with low levels of anxiety? It is possible to divide the subjects into two groups—high anxiety and low anxiety— and ask the question: do these two groups differ in speed of learning? This question must be answered by a significance test.

is not clear, go to Chapters 3 and 12 and peruse the examples illustrating each of these techniques.

INITIAL BRANCHING PROGRAM

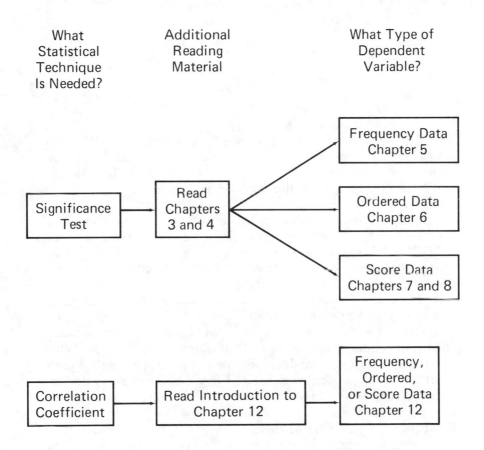

CHAPTER THREE

SIGNIFICANCE TESTS: DETAILED TERMS AND DEFINITIONS

Chapter 2 presented brief definitions of basic terms. It is impossible to overemphasize the importance of understanding these terms. For one thing, it is not possible to understand the process of statistical data analysis without mastering the concepts that underlie these basic terms. In addition, you cannot correctly use the branching programs presented in subsequent chapters unless you can make discriminations based on these terms. Since one of the major purposes of this book is to enable researchers to select the appropriate statistical technique, you must learn the terms so well that you can *always* make correct discriminations.

Since many readers may need more detailed definitions and examples than those presented in Chapter 2, we devote the next two chapters to this need. In this chapter, we discuss material required for the branching program for tests of significance. The researcher must make four essential decisions concerning his research. First: what type of dependent variable do I have? Second: is my design between-subjects, within-subjects, or mixed? Third: how many independent variables do I have? Fourth: how many levels does each independent variable have? In addition to helping the reader make these decisions correctly, we point out exceptions and difficulties that the researcher may encounter. In Chapter 4, we present an example of the use of a statistical test of significance in a rather simple research situation and attempt to clarify such concepts as research, null, and alternative hypotheses, Type I and Type II errors, and alpha and beta levels.

To provide specific practice in using the critical terms, a series of research studies has been presented. Read each of the following examples carefully; they will be referred to throughout the chapter as we illustrate the important concepts involved. The brief name following the number will be used when the study is referred to later.

EXAMPLES

Example 1. Nonsense syllables. Forty subjects learned 12 pairs of non-sense syllables presented visually. Thirty minutes later, they were tested for retention. After an hour's rest they learned a second set of nonsense syllables that was matched with the first set in difficulty. This time, a combined visual and auditory presentation of the stimuli was used. After a 30-minute wait, they were tested for retention on the second set of syllables. In both cases, the researcher measured the number of correct responses.

Example 2. Y maze. Thirty hamsters learned a complex Y maze. In addition, 30 rats were trained on the same maze. For each subject, the researcher measured the number of trials required to reach a criterion of three successive errorless trials.

Example 3. Automobile study. From a list provided by the automobile dealers' association, which lists every new car purchaser for the last six months, a market researcher randomly selected 40 people who had purchased Fords, 40 people who had purchased Chevrolets, 40 people who had purchased Plymouths, and 40 people who had purchased an American Motors car. He asked each person to indicate, on a seven-point scale, how satisfied or dissatisfied he was with his purchase.

Example 4. Conservation. A major conservation bill was to be voted on in the State Legislature. A public-opinion pollster randomly selected 200 Democrats and 200 Republicans and asked them whether they supported or opposed the proposed legislation. He counted the number of Democrats and Republicans who responded each way.

Example 5. Monroe name change. Students at Monroe High School wished to change the full name of the school from James to Marilyn. In order to evaluate the request, school administrators conducted a poll to determine the attitudes toward the proposed name change. Their randomly drawn sample included 100 men, half of them faculty and half of them students, and 100 women, similarly split between students and faculty. Each was asked to indicate whether he favored the present or the proposed name.

Example 6. Case-history diagnosis. All 82 students in an abnormal psychology course were given a test the first day of the term in which they were asked to diagnose various case histories presented in the text. At the end of the term, the 75 remaining students were again given the same test. The instructor measured the number of correct diagnoses each time.

Example 7. Discrimination in monkeys. Eighty monkeys were randomly divided into two groups of 40 monkeys each. One group was kept in a very dimly lighted room for 30 days prior to the experiment. The other group was kept in a normally lighted room for the same length of time. On the day of the experiment, all of the monkeys were required to master a complex discrimination task. Half of the monkeys in each group had been food-deprived for 48 hours prior to the experiment. The other half were fed just prior to the experiment. The experimenter measured the number of correct discriminations made by each monkey.

Example 8. Extroversion and esteem. Three hundred subjects were given an extroversion test. The researcher selected the 30 subjects with the highest scores, the 30 subjects with the lowest scores, and the 30 subjects who were exactly in the middle of the distribution. These 90 subjects were then given a test of their self-concept. All were then subjected to an ego-inflating experience, following which their self-concept was again measured.

Example 9. IQ study. Eighty subjects were given IQ tests. The tests were graded and the subjects were assigned to one of two groups. The 40 subjects with the highest IQ made up one group and the 40 subjects with the lowest IQ made up the second group. All subjects were then asked to assemble a puzzle. The researcher measured the length of time each subject needed to assemble the puzzle.

Example 10. Drugged chimpanzees. Fifteen chimpanzees were each randomly assigned to one of five drug conditions. Three animals received 1 cc of the drug, three received 3 cc, three received 5 cc, three received 7 cc, and three received 9 cc. After the chimpanzees were on the drug for a week, a zoologist who was unaware of the drug schedule ranked the chimpanzees in terms of dominance.

TYPE OF DEPENDENT VARIABLE

The first discrimination required by the branching program is the type of dependent variable used in the study. We mentioned in Chapter 2 that the measurement of a dependent variable generally produces one of three types of data: score data, ordered data, or frequency data. Score data is produced when a numerical score is assigned to the subject's behavior or performance. Ordered data is produced when the subject is ranked along some dimension of interest. Frequency data is produced when the subject is counted as belonging in a certain category or cross-classification of categories.

Let's attempt to apply these concepts to each of our examples. Examples 1, 2, 3, 6, 7, 8, and 9 all represent score data. Examples 4 and 5 represent frequency data. Example 10 is the sole example of ordered data. The emphasis on score data is not accidental: most studies in the behavioral sciences employ score data. Frequency data is encountered less often, and ordered data is probably the rarest type.

In Example 1 (*Nonsense syllables*), the researcher counts the number of correct responses that each subject makes. The fact that he is counting responses *does not* make the data frequency data. The researcher is counting *responses*, not *subjects*. When he finishes counting the responses for each subject, the subject receives this number as his "score," in the same fashion that the number of correct responses on a test may become a subject's score. Each subject's score may be directly compared with every other subject's score. Thus each subject is assigned a numerical score that represents his performance on the task. In Example 2 (*Y maze*), again, the researcher counts the number of trials required to reach a criterion for each animal. This number becomes the score for each animal. In Example 3 (*Automobile study*), each subject's choice of a response, from 1 to 7, becomes his score. In Example 6 (*Case-history diagnosis*), the number of correct diagnoses becomes each subject's score; in Example 7 (*Discrimination in monkeys*), the number of correct discriminations is the subject's score. In Example 8 (*Extroversion and esteem*), the dependent variable is the score that each subject receives on the self-concept test. Finally, in Example 9 (*IQ study*), the length of time needed to assemble the puzzle is each subject's score. We can see from the examples that there are many ways to obtain score data, but all have one thing in common: they all result in a numerical score for each subject that can be compared with all other subjects performing the same task or undergoing the same experience.

Examples 4 and 5 represent frequency data.[1] In these two cases, the researcher is going to count the number of subjects who fall into various categories. In Example 4 (*Conservation*), there are two sets of categories: political party (an individual is either a Democrat or a Republican) and attitude on the legislation (an individual either supports or does not support the legislation). Thus there are four cross-classification categories: Democrats who support the legislation, Democrats who do not support it, Republicans who support it, and Republicans who do not support it. Each subject is counted as falling into one of these four cross-classification categories. Note that no

---

[1] However, see the section *An Ambiguous Case–Dichotomous Data* later in this chapter.

subject is assigned a numerical score. In Example 5 (*Monroe name change*), there are three sets of categories: gender, school status, and opinion on the name change. Thus this example has eight cross-classification categories: faculty men who favor the present name, faculty men who favor the proposed name, faculty women who favor the present name, and so on. Each subject is now counted as falling into one (and only one) of the eight cross-classification categories. Remember that, with frequency data, subjects are counted but no numerical score is assigned to any subject.

Example 10 (*Drugged chimpanzees*) represents ordered data. No numerical score was assigned to the chimpanzees; instead, each of the 15 chimps was ranked on dominance. The most dominant chimp was given the rank of 1, the next most dominant chimp a rank of 2, and the least dominant chimp a rank of 15. Of course, ranking could have proceeded the other way: the most dominant chimp might have been ranked 15, the next most dominant ranked 14, and the least dominant ranked 1. Perhaps it makes more sense to rank the original way, but the statistical analysis yields the same results for either case. One word of caution, however: the researcher must remember how the subjects have been ranked so that there will be no confusion about the meaning of the results when the statistical test is interpreted.

Converting Data

It is important to remember that score data may be converted to ordered data and ordered data may be converted to frequency data, but that the reverse is not true. Frequency data may not be converted to ordered data and ordered data may not be converted to score data. In Example 9 (*IQ study*), there are 40 high-IQ subjects and 40 low-IQ subjects, each of whom has a score that represents the length of time the subject needs to assemble the puzzle. If we wished to, we might take the 80 subjects and rank them from 1 to 80 on the time they required to assemble the puzzle and thus produce ordered data. Or we might arbitrarily call all solutions under three minutes "fast" and all solutions over three minutes "slow". We could then count the number of high-IQ subjects in the "fast" class and the number in the "slow" class and then do the same for the low-IQ subjects, thus producing frequency data.

However, if we have only a table that shows how many high-IQ subjects were "fast" and how many were "slow" and how many low-IQ subjects were "fast" and how many were "slow," we could not then rank the subjects because we would not know how subjects within each classification differed from one another. We would know that all subjects who were "fast" had lower scores than all subjects who were "slow," but we would not know how the "fast" subjects differed from one another or how the "slow" subjects differed from one another. Similarly, we could not reconstruct actual numerical scores from the ranks of the 80 subjects.

Data can be converted in one direction but not in the other because score data generally contain more information than ordered data and ordered data contain more information than frequency data. In other words, assigning a numerical score to represent behavior is usually a more precise form of measurement than assigning a rank.[2] Similarly, assigning a rank is more precise than assigning a subject to a category and counting the number of subjects in each category. When you convert from score data to ordered data or when you convert from score or ordered data to frequency data, you lose information. Such conversions may not always be wise, but they are certainly legitimate. On the other hand, to go from frequency data to ordered data or from frequency or ordered data to score data requires that you obtain information that was not in your initial measurement. Although these procedures might be wise, they are impossible.

Since information is lost in the process, why would one ever wish to convert data to another form? The best answer is that, in some cases, the information that is lost is not believed to be very reliable. If a very crude measuring instrument yields score data, the researcher may feel uneasy about performing an analysis on the scores and may wish to convert the scores to ranks or frequencies. Another reason might be that large quantities of data are involved; in such a case, the analysis might be greatly simplified by setting up a small number of categories and counting the number of subjects in each one rather than analyzing hundreds or thousands of numerical scores. Then, too, in the case of ordered data, only a few statistical procedures are available, and a particular research design may be analyzed more appropriately by a statistical test designed for frequency data. Loss of information, however, also causes a loss in the power of the statistical test to detect real differences between groups. That is, a difference between groups that may be significant when score data are analyzed may not be significant when the data converted to ranks or frequencies are analyzed. Thus data conversion often entails a loss in power for the sake of a gain in convenience.

---

[2] Of course, this statement is true only if the numerical scores are very closely related to the underlying behaviors they represent and have been assigned in such a way that equal differences between scores reflect equal differences in the behaviors (equal interval scale).

An Ambiguous Case—Dichotomous Data

There is one case in which data may be treated either as frequency data or as score data. Whenever a dichotomous dependent variable is employed (that is, when there are only two categories), a score of 1 may be assigned to one category and a score of 0 to the other. In Example 4 (*Conservation*), a score of 1 might be assigned to each person who supported the bill and a score of 0 assigned to each person who opposed it. An analysis designed for score data might then be legitimately applied. Such an analysis would yield virtually the same outcome that would result from counting frequencies in categories and using a test designed for frequency data. In these cases, the researcher is free to use either technique, and his conclusions will be essentially the same regardless of the technique he chooses. When there are more than two categories, however, assigning arbitrary numbers—1, 2, 3—is *not* legitimate. These data may not be treated as score data. They must be treated as frequency data, and the appropriate test for frequency data must be employed. Thus only in the case of dichotomous data is it possible to use these two methods interchangeably.

BETWEEN-SUBJECTS, WITHIN-SUBJECTS, AND MIXED DESIGNS

In a between-subjects design, all comparisons are made between or among different subjects. In a within-subjects design, all comparisons are made within the same group of subjects. Mixed designs incorporate at least one comparison between different subjects and at least one comparison within the same group of subjects. Applying these concepts to our examples, we find that Examples 2, 3, 4, 5, 7, 9, and 10 are between-subjects designs, Examples 1 and 6 are within-subjects designs, and Example 8 is a mixed design.

Between-Subjects

In Example 2 (*Y maze*), the performance of the rats is compared with the performance of the hamsters. In this example, it is quite obvious that different groups of subjects are being compared. In Example 3 (*Automobile study*), the comparison is among people who have purchased Fords, Plymouths, Chevrolets, or American Motors cars. Again it is quite obvious that the comparisons are among different groups of subjects. Example 4 (*Conservation*) also involves comparisons among different groups of subjects. Democrats who support the bill are different people from Democrats who oppose it, and these subjects, in turn, are different from Republicans who either oppose or support it. Similarly, Example 5 (*Monroe name change*) involves comparisons among different groups: one may be either a man or a woman, a teacher or a student, and may favor the present or the proposed name.

Example 7 (*Discrimination in monkeys*) is somewhat more complex, since it involves two comparisons. One comparison is between monkeys who have been kept in the dark and those who have had normal exposure to light. The second comparison is between monkeys who have just eaten and those who have been food-deprived for 48 hours. Because this is a factorially designed experiment, there are four groups of monkeys in all, representing all combinations of rearing conditions and food deprivation. However, each group contains different monkeys, so that the overall design is between-subjects.

Examples 9 and 10 are relatively straightforward. In Example 9 (*IQ study*), high-IQ subjects' performance is compared with low-IQ subjects' performance. In Example 10 (*Drugged chimpanzees*) the comparison is among the five drug-dosage groups, with different chimps serving in the different groups. Hence both of these are between-subjects designs.

Within-Subjects

In Example 1 (*Nonsense syllables*), the performance of the 40 subjects who learned the list of nonsense syllables under the visual presentation is compared with the performance of the same 40 subjects who learned a similar list under a combined visual and auditory presentation. Thus the comparison in performance is made within the same group of subjects who were tested under two different conditions. Similarly, Example 6 (*Case-history diagnosis*) involves a comparison of the performance of the same group of students at the start of the term and at the end of the term. Again, the same group of subjects was measured on more than one occasion.

Mixed Design

Example 8 (*Extroversion and esteem*) is a mixed design. The researcher wishes to compare the performance of those scoring high, medium, and low on extroversion. This comparison is between-subjects, since it involves different groups of subjects. At the same time, the researcher wishes to compare the scores on the self-concept test before the ego-inflating experience with the scores after the experience. This comparison is within-subjects, since the same subjects take the test before and after their ego-inflating experience. Since one comparison (extroversion) is between-subjects and one (self-concept) is within-subjects, this study is a mixed design.

The Importance of the Between- /Within-Subjects Distinction

Many beginning researchers do not understand why it is important to determine whether a research design is between-subjects, within-subjects, or mixed. A simple example may help explain the importance of this distinction.

Let's take the case in which two people have been measured twice on some dimension. On the average, although not always, the two scores from Person 1 and the two scores from Person 2 will be more alike than two scores, one from Person 1 and one from Person 2. For example, the speed at which I learn a list of words will tend to be more similar to the rate at which I learn a second list than it will be to the rate at which *you* learn the *second list*. In short, any two scores derived from my performance will tend to be correlated, while any two scores, one from me and one from you, will tend to be uncorrelated. Statistical tests designed for within-subjects and mixed designs take this correlation into account. If a statistical test designed for between-subjects designs is used to analyze a within-subjects or mixed design, it will (quite possibly) fail to identify a real difference between conditions. Since it is generally desirable to use the most powerful technique available, it is important to be able to determine whether a research design is between-subjects, within-subjects, or mixed. For frequency data, the distinction is even more critical: if data from any subject appears in more than one cell (that is, if there are any within-subject comparisons), the analysis is illegitimate.

Some Sources of Confusion

Determining whether a research design is between-subjects, within-subjects, or mixed is at best confusing and difficult for beginning researchers. You should not be surprised, for example, if you understand the distinctions pretty well today but seem to lose them the next time you attempt to classify a research design. If that happens, go back and reread this section. Eventually the distinctions become relatively permanent. Two situations that are often confusing are discussed briefly here in an effort to aid your discrimination.

Different classes of dependent variables. It is not uncommon for researchers to measure several classes of dependent variables in the same experiment. In Example 3 (*Automobile study*) the researcher, in addition to asking the respondents about their satisfaction with their car, might have asked each of them how much they paid for their car and how many miles they had driven it. Although these questions would yield three measures for each subject—how satisfied he is with his car, how much he paid for it, and how far he has driven it—the result in no way constitutes a within-subjects design. To analyze all three of these scores with a single statistical test would be as inappropriate as the proverbial comparison between apples and oranges, because the three scores represent entirely different classes of dependent variables. In a true within-subjects design, all of the scores may be analyzed with a single statistical test, because the scores represent the same type of behavior or are sufficiently similar to enable the researcher to make meaningful comparisons between the scores. When there is more than one class of dependent variables, several different analyses should be performed. In this case, three separate between-subjects analyses should be performed: one for the satisfaction data, one for the price data, and one for the number of miles driven.

More than one measurement per subject, combined to give one score. It is not uncommon for a researcher initially to take more than one measure on each subject. He may then throw out, average, or combine the scores to yield one final score for each subject. In Example 9 (*IQ study*), each subject might have tried to solve the puzzle three different times. The subject's score might have been the best time of the three attempts, the average of the three, or the combined time for all three. Even though each subject has been measured three times, we do not have a within-subjects design if the analysis is performed on only one score per subject. We have a within-subjects design only if the analysis is performed on more than one score, usually under different conditions, for each subject. Example 9 (*IQ study*) would become a mixed design if all three scores were used separately, with number of trials as a within-subjects variable.

NUMBER OF INDEPENDENT VARIABLES AND NUMBER OF LEVELS OF THE INDEPENDENT VARIABLE

An experiment or research project can assess the effects of one or more independent variables on the dependent variable. If more than one independent variable is employed, variables are usually combined factorially—that is, each level of one independent variable is combined with each level of every other independent variable. (In a factorial experiment, the number of different experimental conditions equals the number of levels of Independent Variable 1 times the number of levels of Independent Variable 2 times the number of levels of Independent Variable 3, and so on. Thus if one independent variable has four levels and a second independent variable in the same experiment has three levels, there will be 12 experimental conditions in all. The 12 conditions will exhaust all of the possible combinations of levels of the two independent variables.)

Let's identify the number of independent variables and the number of levels of each independent variable for our ten examples. There is one independent variable in Example 1 (*Nonsense syllables*): method of presentation. There are two levels: visual and visual plus auditory. Example 2 (*Y maze*) has one independent variable—species of animal—with two levels: hamster and rat. Example 3 (*Automobile study*) has one independent variable—type of car owned—with four levels: Ford, Plymouth, Chevrolet, and American Motors.

For the moment, we shall pass over Examples 4 and 5. Frequency data present unique problems of identifying the number of independent variables. We shall return to this problem after we have surveyed the remaining five examples.

Example 6 (*Case-history diagnosis*) has one independent variable —time of administration of test—with two levels: start of term and end of term. There are two factorially combined independent variables in Example 7 (*Discrimination in monkeys*). One independent variable is amount of illumination in the vivarium, with two levels: normal and dim. The second independent variable, number of hours of food deprivation, also has two levels: 0 hours and 48 hours. Since each independent variable has two levels, there are four experimental conditions: dim illumination, 0 hours' deprivation; dim illumination, 48 hours' deprivation; normal illumination, 0 hours' deprivation; and normal illumination, 48 hours' deprivation. Example 8 (*Extroversion and esteem*) also involves the factorial combination of two independent variables. The first is degree of extroversion, with three levels: high, medium, and low. The second is time of administration of the self-concept test, with two levels: before and after the ego-inflating experience. Since one independent variable has three levels and the other has two, there are six experimental conditions: high, before experience; high, after experience; medium, before experience; medium, after experience; low, before experience; and low, after experience.

Example 9 (*IQ study*) has one independent variable, IQ, with two levels: high and low. Example 10 (*Drugged chimpanzees*) has one independent variable, amount of drug dosage, with five levels: 1 cc, 3 cc, 5 cc, 7 cc, and 9 cc.

The Strange Case of Frequency Data

Let's return to Examples 4 and 5. You will remember that earlier in this chapter we noted that *if the response categories in frequency data are dichotomous*, a score of 1 or 0 may be assigned to the responses and the dependent variable may be treated as score data. If we assign these values in Example 4 (*Conservation*), we find that we have one independent variable, political party affiliation, with two levels: Democrat and Republican. The score of 0 or 1, which represents support for or opposition to the bill, becomes the dependent variable. If we assign scores of 0 or 1 to preference for present or proposed name in Example 5 (*Monroe name change*), we have two independent variables that are factorially combined. One independent variable is gender, with two levels: men and women. The second independent variable is school status, with two levels: faculty and student.

However, if we treat the data as frequency data, we must change our perspective. Example 4 now must be considered to have two independent variables instead of one. One independent variable, again, is political party affiliation, with two levels: Democrat and Republican. The second independent variable is response to the bill, with two levels: support and opposition. The dependent variable now becomes the frequency of people who fall into each cross-classification of these categories. Similarly, if we treat the data of Example 5 as frequency data, we have three independent variables instead of only two. The independent variables are gender, school status, and opinion on name change. Each of these independent variables has two levels: man and woman, faculty and student, and preference for present name or preference for proposed name.

When frequency data are analyzed, the variable that would be the dependent variable if we were using score data must be treated as an independent variable. Thus if you were to treat the data of Example 4 as frequency data, you would look in the branching program for a statistical test for frequency data designed to handle two independent variables. If you assigned scores of 1 or 0 to the responses (that is, if you treated the data as score data), you would look in the branching program for a statistical test for score data designed to handle one independent variable. Similarly, Example 5 would be referenced in the branching program as having three independent variables if the data were treated as frequency data and as having two independent variables if the data were treated as score data.

NOTE: The astute reader may have noticed that in some ways the two examples of within-subjects design, Examples 1 (*Nonsense syllables*) and 6 (*Case-history diagnosis*), represent poor research designs. In general, within-subjects designs pose problems that are not encountered in between-subjects designs. In both the examples, changes from Measurement 1 to Measurement 2 could be attributed to factors other than the independent variable. In Example 1, if subjects learned better under the combined audiovisual presentation than under the visual-only presentation, the researcher would like to attribute the difference to the method of presentation. However, it is possible that subjects learned better the second time because learning the first task made it easier to learn the second task, because they were more relaxed and less tense the second time, and so on. If the results indicated that they learned better under the visual-only presentation, it could be because they were fatigued the second time, because they were bored, or because learning the first task interfered with their ability to learn the second task. In other words, order of presentation (first or second) is confounded with method of presentation (visual or audiovisual), the variable with which the experimenter is actually concerned.

Most of the problems in Example 1 can be eliminated by the simple process of counterbalancing. That is, half of the subjects might be given visual first and audiovisual second and the other half audiovisual first and visual second. In this way, fatigue, boredom, warm-up, and retroactive or proactive inhibition effects are balanced out of the design. We did not present the example with counterbalanced treatments because there are two distinct ways to analyze the data in a counterbalanced experiment. First, we might simply ignore the counterbalancing and analyze the design as having one independent variable with two levels. This is a perfectly legitimate technique if the researcher is not interested in order effects but merely wants to balance them out in the experiment.

However, the researcher might test for the existence of order effects in the study. The design then would have two factorially combined independent variables (order of presentation, method of presentation), each with two levels. He would then have a mixed design rather than a within-subjects design. Method of presentation (audiovisual and visual) still constitutes a within-subjects factor, since each subject serves in both conditions. Order of presentation (visual first or audiovisual first) is a between-subjects factor. One group of subjects gets audiovisual first and visual second; a different group of subjects gets visual first and audiovisual second. Thus the same study is analyzed as a within-subjects design if the counterbalancing is ignored and as a mixed design if the counterbalancing is tested.

Example 6 (*Case-history diagnosis*) presents a similar problem. If the students average more correct diagnoses at the end of the semester than at the beginning, it is not clear that taking the course is the reason for the improvement. The students are older than at the start of the semester, they have taken other college courses, they have interacted with more people, and so on. In this case, you cannot eliminate these other explanations by counterbalancing. There is obviously no way that half of the students can take the test first at the end of the semester and then take it again at the start of the semester when they are five months younger. It would be very difficult to design a study that would yield clear information about the effect of the course. We cannot include a full discussion of methods of handling these problems; however, we strongly recommend that the researcher who plans to use within-subjects designs read the discussion of these issues by Campbell and Stanley (1966). These authors devote a great deal of attention to experimental designs with more than one measure per subject.

If within-subjects experiments are so difficult to design and interpret, the reader may well wonder why they are used at all. There are several reasons. Properly designed, they are more precise and powerful than between-subjects designs that utilize the same number of subjects. As a corollary, experiments of equal precision and power can be run using far fewer subjects than are necessary for a comparable between-subjects design. Finally, some research problems, such as discrimination reversals and learning sets, require within-subjects designs. However, the researcher who is using a within-subjects design should always scrutinize his experiment carefully to ensure that all confounded variables and competing hypotheses have been ruled out.

SUMMARY AND EXERCISE

At this point, we have identified all of the relevant characteristics of the ten research designs presented at the beginning of the chapter. While the material is still fresh in your mind, go back to each example and do the following exercise:

1. Determine the type of dependent variable.
2. Decide whether the design is between-subjects, within-subjects, or mixed.
3. Determine the number of independent variables.
4. Determine how many levels each independent variable has.
5. Begin with the branching program at the end of Chapter 2 and continue until you have identified the appropriate statistical test for each example. If you have correctly identified the relevant dimensions and the appropriate statistical test, you are ready to use the branching program. If you have made any mistakes, reread Chapters 2 and 3 until you can perform the exercise with 100 percent accuracy.

ANSWERS TO THE EXERCISE

Example 1: *Nonsense syllables*

1. Score data
2. Within-subjects

3. One independent variable
4. Two levels
5. One-way $\times$ subjects ($2 \times s$) ANOVA; t test for correlated means

Example 2: *Y maze*

1. Score data
2. Between-subjects
3. One independent variable
4. Two levels
5. One-way ANOVA; t test

Example 3: *Automobile study*

1. Score data
2. Between-subjects
3. One independent variable
4. Four levels
5. One-way ANOVA

Example 4: *Conservation study*

1. Frequency data
2. Between-subjects
3. Two independent variables
4. Two levels of each
5. Two-way chi square ($2 \times 2 \chi^2$)
   or
1. Score data
2. Between-subjects
3. One independent variable
4. Two levels
5. One-way ANOVA; t test

Example 5: *Monroe name change*

1. Frequency data
2. Between-subjects
3. Three independent variables
4. Two levels of each
5. Three-way chi square ($2 \times 2 \times 2 \chi^2$)
   or
1. Score data
2. Between-subjects
3. Two independent variables
4. Two levels of each
5. Two-way (2×2) ANOVA

Example 6: *Case-history diagnosis*

1. Score data
2. Within-subjects
3. One independent variable
4. Two levels
5. One-way × subjects (2 × s) ANOVA; t test for correlated means

Example 7: *Discrimination in monkeys*

1. Score data
2. Between-subjects
3. Two independent variables
4. Two levels of each
5. Two-way (2 × 2) ANOVA

Example 8: *Extroversion and esteem*

1. Score data
2. Mixed design
3. Two independent variables
4. Three levels of one, two levels of the other
5. Between-within (2 × 3) ANOVA

Example 9: *IQ study*

1. Score data
2. Between-subjects
3. One independent variable
4. Two levels
5. One-way ANOVA; t test

Example 10: *Drugged chimpanzees*

1. Ordered data
2. Between-subjects
3. One independent variable
4. Five levels
5. Kruskal-Wallis test

CHAPTER FOUR

THE NATURE OF A STATISTICAL TEST OF SIGNIFICANCE

As you are wending your weary way to a day's work at the Snerd Factory, a shady-looking character jumps out of the bushes and accosts you with the following proposition: "Ain't you one of the guys who stops off at Maxie's Bar and Grill after work and flips coins with the other guys to see who's going to buy the beer that afternoon?" You warily nod assent. He continues, "I got a real deal for you. A buddy of mine makes up these here loaded coins by splitting them in half, drilling some of the metal out, and then gluing them back together. When you flip them, they come up either heads or tails most of the time, depending on which side he drills the metal off. I'll sell you this here dime for only ten bucks. Believe me, you'll make that ten bucks back in two weeks with all the pitchers of beer you'll win."

The idea immediately intrigues you—you really could win a bundle with a loaded coin. You examine the coin closely and can see no sign that it has been doctored. You ask the character, "Does it come up heads more often or tails more often?" "Geez, buddy, I dunno," he replies. "He makes both and I don't remember what this one is." Now you're really in a quandary. If the coin really is loaded, it is worth ten dollars to you. But how do you know that anything has really been done to it? The appearance of the character doesn't do much to reassure you. Maybe he's trying to sell you a perfectly ordinary dime for ten dollars! You ask him, "How do I know that anything has really been done to it?" He says, "Flip it ten times and then decide but hurry up—I ain't got all day."

At this point, you might larcenously consider, "If the coin is loaded, it's worth ten dollars to me. I'd hate to not buy it if he's telling the truth. On the other hand, I'd feel like a fool if I paid him ten dollars for an ordinary dime. When I flip it ten times, if it's an ordinary dime, the most usual outcome would be five heads and five tails. But even if it is fair, I'd expect that sometimes it would come out four heads and six tails or six heads and four tails. Occasionally it would come out seven and three, but that would be rarer than six and four. I would even expect that it could come out eight and two, nine and one, or even ten and zero, but each of these outcomes would be rarer than the preceding one. I guess I will just have to decide on some particular outcome and buy it if it does that and refuse to buy it if it doesn't. I think I'll buy it if it turns up at least eight heads or tails in the ten flips."

You take the coin and flip it ten times. It comes up nine heads and one tail. You pay the character his ten dollars and put the dime in your pocket. As you continue your walk to the Snerd Factory, you realize that there is some possibility that you in fact paid ten dollars for an ordinary dime, but that the odds are in your favor that it is really loaded.

This frivolous example may appear unrelated to research or statistical tests of significance, but in fact it encapsulates many aspects of the decision-making process with which the researcher must struggle. First, you must generalize from a sample (ten flips) to a population of interest (all of the flips the coin could make). Second, you must make a decision (to buy or not to buy) based on incomplete information (the outcome of your ten flips). Third, there is some information available about expected outcomes if in fact there was no treatment effect (the coin is fair). In this example, the probability that a fair coin would turn up eight heads and two tails or two heads and eight tails in ten flips may be determined precisely. A split at least as extreme as eight heads and two tails (or the reverse) would be expected with a fair coin only about 10% of the time. Fourth, there is the problem of making a mistake: either buying an unloaded coin or failing to buy a loaded coin. In both cases there is some probability that any decision is the wrong decision.

Now let's present these concepts more formally in an example cast in a research setting. Mr. Hildebrant has developed a new method of teaching sixth-grade mathematics, using a teaching machine that employs his own "quadrilateral branching program." He believes that his "quadrilateral branching program" is superior to any known method of teaching mathematics, but he is realistic enough to know that it might not actually be superior and that it might even be inferior to the present method. The school board in his district is interested in his new method but is concerned about the expense of implementing it. However, they would change methods if they were reasonably convinced of its superiority. To help them make their decision, they give Mr. Hildebrant a grant to purchase 16 machines. He must select 16 students at random from the sixth grades in the school system and teach them using the machines and his new program. At the end of the school year, his 16 students are given a standardized test that has been used in the district for years. The mean of the standardized test is 200, the standard deviation is 32, and the scores are normally distributed. Figure 4.1 shows the distribution of test scores for the standardization group.

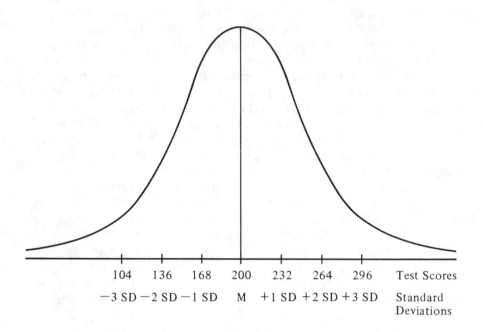

	104	136	168	200	232	264	296	Test Scores
	−3 SD	−2 SD	−1 SD	M	+1 SD	+2 SD	+3 SD	Standard Deviations

Figure 4.1. *The distribution of test scores for the standardization group.*

Mr. Hildebrant encounters the same problems you found in the coin-flipping example. He must generalize to a population of interest (all potential sixth-grade math students) from the results of a sample (16 randomly selected sixth-grade math students). He must decide whether or not to urge the school board to convert to his system on the basis of information from his small sample. He knows what to expect if his method is not superior (his students should average about 200 on the standardized test). He knows that there are two possible mistakes he can make: he can falsely conclude that his method is different when it really isn't or he can falsely conclude that his method is not different when it really is. Finally, whichever decision he makes, he knows that there is some possibility that it's wrong.

Let's see how he handles his problem. One of the first things he must do is obtain more precise information about how his students may be expected to perform if his method is no different from previous methods. They should average about 200 on the test, but it is unlikely that a group of 16 would score exactly at the mean of the larger standardization group. Just as the coin-flipper realizes that a fair coin will come up seven heads and three tails a certain percentage of the time, Mr. Hildebrant knows that even if his technique does not differ in effectiveness from previous techniques, his group of 16 students might average 180, or 196, or 210 some of the time. To obtain accurate probability estimates of these outcomes, he must know how the average scores for randomly chosen groups of 16, called "groups of size 16," are likely to distribute themselves.

Distribution of sample means. He can obtain this information by performing some mathematical transformations on the distribution of scores (see Figure 4.1), which indicates how *single individuals* can be expected to perform. To determine average performance of groups of size 16, Mr. Hildebrant might draw random samples of size 16 from the population of scores, record the mean for each group, and develop a new distribution. This new distribution, the distribution of sample means[1] for samples of size 16, would provide some information about the expected performance of groups of size 16. Fortunately, we need not go through the time-consuming and tedious process of drawing thousands of random samples and developing a new distribution. Statisticians have proved that, if a very large number of samples is drawn, the mean of a distribution of sample means ($M_{\bar{X}}$) equals the mean of the population from which it is derived (μ). Further, the standard deviation of a distribution of sample means ($\sigma_{\bar{X}}$) equals the standard deviation of the population (σ) divided by the square root of the sample size (N). In other words, $M_{\bar{X}} = \mu$ and $\sigma_{\bar{X}} = \sigma/\sqrt{N}$. In this case, the mean of the distribution of sample means equals 200 and the standard deviation of the distribution is obtained by dividing 32 (the standard deviation of the population) by the square root of

[1] The "distribution of sample means" is ordinarily referred to as the "sampling distribution of means," but we feel that this label is confusing and sometimes misleading for the beginner.

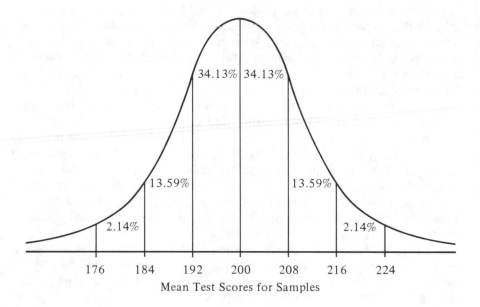

Figure 4.2. *Distribution of sample means for samples of size 16, derived from the population of scores shown in Figure 4.1.*

the sample size: $32/\sqrt{16} = 8$. The distribution of sample means is shown in Figure 4.2.

As can be seen, the distribution of sample means is also normally distributed. Using his knowledge of the normal curve,[2] Mr. Hildebrant now knows that if his new technique is no different from the older techniques, his sample of 16 would be more likely to score 200 than any other single score. But he also knows how often other scores would be expected *even if the new technique made absolutely no difference in learning* as compared to the old techniques used when the test was standardized. Thus he knows that about 32% of the time they would average higher than 208 or lower than 192, that about 5% of the time they would average higher than 216 or lower than 184, and that less than 1% of the time they would average more than 224 or less than 176.

Mr. Hildebrant's logic now proceeds as follows: "If I assume that my new method is not different from the old method, I can use the

---

[2]The normal curve provides information about the percentage of cases that lie between different scores for any normally distributed data with a known mean and standard deviation.

distribution of sample means to estimate the probability that certain outcomes will occur. For example, suppose I obtain an average over 216 or under 184. This should happen only 5% of the time *if* my method does not differ from the old method. Thus if I decide to reject the null hypothesis should I obtain an average over 216 or under 184, then I have a 5% chance of falsely concluding that the new method is different from the old when there is in fact no difference.[3] However, to avoid making a mistake, I should conclude that the new technique is different from the old only if my sample averages over 224 or under 176, because then I would have only a 1% chance of mistakenly concluding that the new technique is different when it really isn't."

Null hypothesis, alternative hypothesis, and research hypothesis. There is a problem at the end of Mr. Hildebrant's logic, but let's ignore it for a moment while we examine his other considerations. His research hypothesis is that the students who learn mathematics from his program will perform better, on the average, than students who have learned math under conventional teaching methods. His two-tailed null hypothesis is that his sample of 16 students were drawn from a population with $\mu = 200$. Note that this is just a formal way of stating that his new method is no different from the old method, for if his new method has no effect he would expect his students to perform similarly to students who had learned under the old method and had averaged 200 on the standardized test.

The alternative hypothesis would be that his sample of 16 students was drawn from a population with $\mu \neq 200$.[4] This alternative hypo-

---

[3] Technically, what Mr. Hildebrant means by this statement is, "If my method does not work (the H_0 is true), and I were to replicate the study a large number of times, I would mistakenly conclude that it works only 5% of the time."

[4] For simplicity in the following section and in the remainder of this book, we consider only one of several possible situations: namely, that the null hypothesis is that the treatments have yielded the same value ($H_0 : \mu_1 = \mu_2 \ldots$). We therefore consider only a single alternative hypothesis (although there are as large a variety of these as there are null hypotheses): namely, $H_1 : \mu_1 \neq \mu_2 \ldots$. We have chosen to do this because the great majority of applications, especially those encountered by students, may be appropriately handled by this simplest case, and because it is beyond the scope of this book to consider the exhaustive variety of other alternatives. Other alternatives include (1) one mean is greater than or equal to another ($H_0 : \mu_1 \geq \mu_2$); the alternative hypothesis is that the mean is less than the other ($H_1 : \mu_1 < \mu_2$). (2) A mean equals a specific value ($H_0 : \mu_1 = 10$). The alternative hypothesis is that the mean does not equal the specific value ($H_1 : \mu_1 \neq 10$).

You should remember that there are other null and alternative hypotheses. If the simple hypothesis is not adequate for your purposes, consider one of these. For further information on developing statistical hypotheses, see Kirk (1968), pp. 22–25.

thesis is broader than the research hypothesis because it includes the possibility that his students may perform less well than previous students. Therefore, rejection of the null hypothesis and acceptance of this alternative hypothesis does not provide automatic support for the research hypothesis. The research hypothesis would be considered supported only if the null hypothesis were rejected *and* the mean of his group of 16 students was in fact higher than 200.

His decision on whether to reject or accept his null hypothesis is based on the actual performance of the students in his sample. Therefore, if he obtained a mean for his sample as high or higher than 224 or as low or lower than 176 (these results would occur less than 1% of the time if the null hypothesis were true), he would reject his null hypothesis in favor of his alternative hypothesis. In so doing, he runs a 1% risk of concluding that his technique has had an effect when it actually has not. In effect, the risk he runs is that he might reject a true null hypothesis.

Type I and Type II errors. This mistake, rejecting a true null hypothesis, is known as a Type I error. The probability of making a Type I error is called alpha (α). It is also possible to make an error by failing to reject a false null hypothesis. This error is known as a Type II error. The probability of making a Type II error is called beta (β). In any experiment, the null hypothesis is either true or false. The coin is fair or it is biased (loaded). The technique doesn't work or it does. Every researcher wishes to reject false null hypotheses and accept true ones. This situation is diagrammed in two different ways in Figure 4.3. In Mr. Hildebrant's case, a true null hypothesis means that for the total population, his new technique does not differ from the older techniques. A false null hypothesis means that it does differ. If Mr. Hildebrant makes a Type 1 error, he (falsely) concludes that the new technique differs when it does not. If he makes a Type II error, he (falsely) concludes that the new technique does not differ when it does.

Let's return now to examine that earlier flaw in Mr. Hildebrant's reasoning. He had decided that he would conclude that his program made no difference unless his group averaged over 224 or under 176 (α = 1%) rather than over 216 or under 184 (α = 5%). He reasoned that a more stringent criterion would prevent his concluding that the technique worked when it didn't, thus reducing his chance of making a Type I error. Unfortunately, all other things being equal, the more protection you have from a Type I error, the more likely is a Type II error. We will try to clarify this point with another illustration.

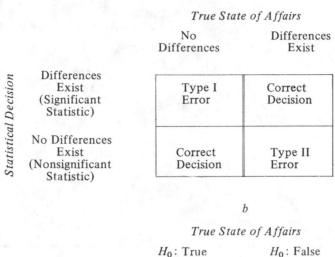

Figure 4.3. *The relationship of researcher decisions and true or false null hypothesis to Type I and Type II errors.*

Let us assume for a moment that Mr. Hildebrant's new technique really does work, and that, furthermore, it will add an average of 16 points to the score of every student who is taught by the new method. (Although we will make this assumption for this illustration, in a real experiment, of course, this is something we would never know.)

In Figure 4.4 we present the distribution of sample means for samples of size 16 drawn from a population with a mean of 200. Superimposed on this figure is another distribution of sample means, this one for samples of size 16 drawn from a population with a mean of 216. The population with the mean of 216 represents the way in which future students would score if they averaged 16 points better after having been taught by the new method. Mr. Hildebrant has set his α level, and thus the probability of a Type I error, at 5%. That

Distribution Distribution
     1   2

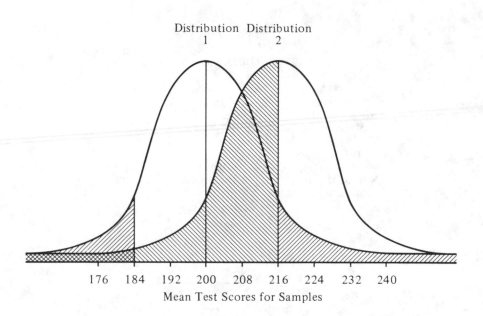

176   184   192   200   208   216   224   232   240

Mean Test Scores for Samples

Figure 4.4. *Distribution of sample means for H_0 : True (Distribution 1) and H_0 : False, M = 216 (Distribution 2). Area of Distribution 1 that has diagonal lines represents the probability of Type I error; area of Distribution 2 with diagonal lines represents the probability of Type II error.*

is, he will reject his null hypothesis if his students score higher than 216 or lower than 184. Using the superimposed distribution of sample means, we can determine his probability of making a Type II error. If his students really come from a population with a mean of 216, there is a 50% chance that they will score lower than 216 (half of the scores in a normal distribution are below the mean). If their mean is lower than 216, he cannot reject his null hypothesis, and he will therefore make a Type II error. Since Mr. Hildebrant does not know that his new technique will add an average of 16 points to each student's score, he does not know that he has a 50% chance of making a Type II error. But we do, and 50% seems a very high risk of deciding falsely that the new technique did not work.

What happens if Mr. Hildebrant lowers his risk of a Type I error by setting his alpha level at .01, rather than .05? In that case, in order to reject the null hypothesis, his sample would have to average above

224 or below 176. For the superimposed distribution of sample means, 84% of the scores lie below 224, since it is one standard deviation above the mean in that distribution. Now, Mr. Hildebrant's risk of committing a Type II error has risen to 84%!

Controlling Type II errors. In the real world, unfortunately, it is very difficult to estimate the probability of a Type II error except in the rare case when you have some information about the size of the effect of the experimental treatment. However, some general rules can help minimize your Type II error rate. You have already seen one of these rules: since you, the experimenter, choose your own Type I error rate, the more stringently you try to protect yourself from falsely rejecting a true null hypothesis, the more likely you are to accept the null hypothesis if it is false. Therefore, the most lenient Type I error rate that is compatible with the purposes of the experiment will provide the best insurance against a Type II error. We will have more to say about this problem in a later paragraph.

There are three other strategies that minimize the probability of a Type II error. The first is to *increase the size of the sample if possible*. If Mr. Hildebrant had been able to obtain 64 teaching machines instead of only 16, or had he been able to teach four students with each machine, he could have obtained a random sample of 64 students. Our distribution of sample means has a mean of 200 for samples of size 64 just as for samples of size 16. However, the standard error (the standard deviation of the distribution of sample means) for samples of size 64 equals $32/\sqrt{64} = 4$, as compared with 8 for samples of size 16. In order for him to conclude that there are differences with an alpha level of about 5%, his sample of 64 students must score above 208 or below 192, rather than above 216 or below 184, as was the case with a sample of 16. If his new technique increased scores an average of 16 points, the probability that the average of the group of size 64 would fall below 208 and that he would fail to reject the null hypothesis is now less than $2\frac{1}{2}\%$. Therefore, with a sample of 16 students, an alpha rate of 5%, and an average effect of 16 points, the probability of a Type II error was 50%, as we have seen. With the same alpha level, the same effect, but a sample of 64 students, the probability of a Type II error is only $2\frac{1}{2}\%$. Increasing the size of the sample is always an effective way to reduce the probability of a Type II error. (You may find it a useful exercise to draw the distributions for sample size 64.)

If the study is a true experiment, the second strategy is to *increase the size of the effect*. If Mr. Hildebrant's branching program increased performance by 32 points rather than by 16, there would

be much less chance of a Type II error. Similarly, if the program increased performance by 100 points, the possibility of a Type II error
would be negligible. The more Mr. Hildebrant can refine and perfect
his program to have a maximal effect on performance, the more he
reduces the probability of a Type II error.

The third strategy is to *reduce the extraneous variability*. If Mr.
Hildebrant had selected subjects with a narrower range of abilities,
had provided uniform sets of instructions, and so on, he would have
reduced extraneous variability and would have been more likely to
find differences.

Let's briefly summarize the material on Type I and Type II
errors. In any given piece of research, either the null hypothesis is
true (there is no relationship between the independent and dependent
variables) or the null hypothesis is false (there is a relationship). If
the null hypothesis is true, you may either make a correct decision
and accept it or you may make a Type I error and reject it. If the
null hypothesis is false, you may either make a correct decision and
reject it or you may make a Type II error and accept it. However,
since you do not know whether the null hypothesis is true or false,
you must protect yourself against the possibility of both types of
errors. You determine the probability of a Type I error by selecting
a level of alpha. The probability of a Type II error is influenced by
the alpha level, the size of the sample, and the strength of the
experimental effect (which is rarely known in advance). It can be
minimized by setting a low alpha level, using the largest samples
possible, choosing the levels of the independent variable that have
the greatest effect upon the observed behavior, and minimizing
extraneous variability.

Selecting your alpha level. What is the appropriate alpha level?
Historically, researchers in the behavioral sciences have selected 5%
as the highest risk of a Type I error that they are willing to run, although many have argued that 1% is a more appropriate level. However, there is nothing sacred about either of these levels. In many cases,
the alpha level should be set by a consideration of the consequences
by making a Type I as opposed to a Type II error. In our example,
the consequences of making a Type I error are that the school board
would spend a great deal of money converting to teaching machines
when the machines would not lead to better learning than the older
methods. The consequences of a Type II error are that Mr. Hildebrant
would not suggest conversion to the teaching machines and that
future math students would be denied a teaching technique that
would increase their mathematical skills. Probably only Mr. Hildebrant

and the school board can evaluate how disastrous each of these mistakes could be. If they were terribly concerned about the expense of changing systems and not terribly concerned about denying future generations of students a new and improved technique, they should set a rather stringent alpha level, such as .01. If the money were relatively unimportant to them and they were very concerned about the possibility of not using the most efficient teaching techniques available, an alpha level of .05, .10, or even .20 might be a better choice.

In much behavioral science research, we are concerned with allocating our energies to one kind of problem or another. On the one hand, we do not wish to invest too much energy in an unpromising area; and on the other, we do not wish to overlook an area that might yield interesting relationships. In these cases, the .05 level is accepted as the most appropriate choice. Because it has historical justification, it is accepted by editors of scientific journals as an appropriate alpha risk. At the same time, it minimizes the risk of a Type II error better than more stringent levels.

One final point must be made about alpha level. Every statistics student learns that the researcher should choose the alpha level in advance, before the analysis is performed. But journals are filled with reports of findings significant at the .01, the .001, or even more extreme levels. Does this mean that the scientist has adopted such a stringent criterion of statistical significance? Although there is some disagreement on and misunderstanding about the meaning of the values reported, these reported values are generally *the probabilities associated with the test statistic* rather than the alpha level established by the scientist. They are provided by the writer for the convenience of the reader, giving information that the reader would otherwise have to obtain from a table.

One must be cautious, however, in interpreting these values. Most students feel that a difference significant at the .001 level is "better" than a difference significant at only the .05 level. And some people report the most stringent level of significance because they believe that it indicates how strong the relationship is. A better indication of a large and significant difference is a strength-of-association measure (see Chapter 11), which is designed specifically to tell you how strong a relationship is.

Although you will report the probability associated with the test statistic in your report of your findings, you should remember that you are protected against a Type I error not at these particular reported levels but at the significance level that you have established, usually $\alpha = .05$.

One- and two-tailed tests. Another problem area lies in the decision between the one-tailed and the two-tailed test. In our example, Mr. Hildebrant was interested in learning whether his students performed less well or better than the previous students, even though he hoped they would perform better. As a consequence, for $\alpha = .05$, he decided to reject his null hypothesis if his students' performance fell in the upper $2\frac{1}{2}\%$ or in the lower $2\frac{1}{2}\%$ of the distribution of sample means. However, had he wished to evaluate only the possibility that his students would perform better, he could have eliminated the critical $2\frac{1}{2}\%$ in the lower half of the distribution and added it to the $2\frac{1}{2}\%$ in the upper half, thus putting all of his critical area for rejecting the null hypothesis in the upper half of the distribution. The net effect of this change would have been that, with 16 subjects, an average performance exceeding only about 213.2 points, instead of 216 points, would have led him to reject his null hypothesis. He could thus have run a one-tailed test instead of a two-tailed test and increased his chances of rejecting his null hypothesis.

There is continued controversy about the appropriateness of using the one-tailed test. There is agreement on one point, however: namely, that the decision to perform a one-tailed test must be made at the time the statistical hypothesis is formulated and before the test of significance is performed. If you make the decision after an initial test has indicated that the two-tailed statistic is not significant, the alpha level you are setting is considerably less stringent than it appears.[5] We recommend that you avoid using one-tailed statistics until you have read sufficiently on the issue to understand both the pitfalls and the advantages of these procedures.

A NOTE ON POPULATION

Most people readily understand the notion of *sampling*, but many appear to have some difficulty understanding the notion of *population*. A population is the total group, real or hypothetical, from which our sample is drawn and to which statistical inferences may be made on the basis of the results of our research. Statistical inferences can be made to populations only if the samples involved are randomly drawn from the populations. If subjects are not randomly drawn from a population—for example, if we take the first ten volunteers from an introductory psychology class, or if we employ the ten animals in our laboratory—nonstatistical inferences to appropriate populations may be made providing there is random assignment of these subjects

[5]For a variety of views on this issue, see Kirk (1972), Chapter 8.

to conditions.[6] Since most samples are not randomly drawn from populations, we must be particularly concerned about the populations to which we make inferences.

Suppose, for example, that we perform a study on *a sample of students from the introductory psychology class at a California state university*. If we find a significant difference in the performance of our sample under two conditions, to what population may we generalize our conclusion that such differences exist? The class of variables we are studying are important to our logical considerations. If the variables are almost unaffected by the physical or social environment, then we may draw conclusions for quite broad populations. If they are affected by these factors, however, we must be very careful in extending our inferences broadly. Here we present an abbreviated list of logical considerations that may limit the breadth of the population to which a researcher may generalize on the basis of his study. This list ranks the considerations from a low to a high order of generalization.

1. The finding is true of our sample (and we have evidence that the sample result could be expected to occur again). This statement, of course, does not involve generalization *from* this sample to a population, and statements about the sample are ordinarily not particularly interesting to the researcher.

2. May we generalize our findings to *all introductory psychology students* taking the course this semester at the university? Is our sample *randomly* drawn from this more general population? If it is, we may make an inference to this population. If it is not, then we must consider whether our relationship is one that may be affected by the social or physical environment. If it is, then, before we may generalize (logically) to this larger population, we must ask ourselves certain questions: Has our sample systematically excluded some volunteer subjects? Have fewer men than women signed up for the experiment? If so, have men been systematically excluded? Might men and women be expected to perform differently in this experiment? Was the study run early in the term, when early-bird go-getters might be over-represented? Was the study run late in the term, when procrastinators or those who hate serving in experiments might be over-represented? Are the variables in our study ones that might be sensitive to differences in motivation? Only if we are reasonably certain that our sample logically represents the population taking introductory psychology this semester may we appropriately generalize to this population.

---

[6] For an excellent discussion of this topic, see Edgington (1966).

3. May we generalize our findings to *all students at the university*? (We know that we do not have a random sample and we must therefore depend on logical inference.) Has the fact that the course has been changed from a requirement to an elective changed the representativeness of the students taking the class? Is a sample from this elective class in fact representative of the total university population? If students from the sciences or the humanities systematically choose other electives, the population from which our sample is drawn may be quite different from the university population.

4. May we generalize our findings to *all university students*, regardless of the university they attend? Our logical considerations must include regional differences, the differences in the kinds of students that might attend our university and other universities, and the likelihood that our variable might be sensitive to such differences in populations.

5. May we generalize our findings to *all people*? In addition to our earlier questions, is a group of 17- to 21-year-old students in fact representative of people of all ages? Is a population of college-educated (or college-eligible) students representative of all people regardless of educational attainment?

For most variables, we would rarely be willing to make the sweeping generalizations suggested in number 5 on the basis of a single study or on the basis of a series of studies done on samples drawn from similar subpopulations. On the other hand, we are usually unwilling to restrict our statements simply to our sample or to make the very low-order generalizations such as number 2. Generalizations are most often made to some intermediate population, such as to the population of students that attends (or might attend) our university or universities similar to our own. To generalize to this population means we are willing to logically defend our sample as being representative of this more general population. The scope of the generalization must often be adjusted to the kind of variable we are investigating. Some variables are likely, and others unlikely, to be affected by gender or by regional, educational, social, physical, or other differences.

A NOTE ON THE EXAMPLE

Keep in mind that the type of example we presented, drawing a sample from a population with known parameters, is encountered very rarely in actual research. It was chosen primarily for pedagogic reasons. In actual practice, Mr. Hildebrant probably would have randomly selected 16 students to teach with his branching program and randomly selected 16 other students (a control group) to teach by the old method. His null hypothesis would have been that the two samples were drawn from populations with equal means, rather than that his sample was drawn from a population with a mean equal to 200. His statistical test would have compared the difference between the means of his two samples. We chose to illustrate the logic of a significance test with the example of a single sample drawn from a population with known parameters because it is somewhat easier to illustrate some important concepts with such an example. However, all statistical tests involve basically the same steps and are subject to the same problems.

SUMMARY

The rather simple example we have discussed illustrates basic features common to all statistical tests of significance. In all cases, a null hypothesis is formed. The exact form of the null hypothesis depends on the type of data and the nature of the research problem. The null hypothesis is some statement about population values that, if rejected, allows for the possible acceptance of the research hypothesis. A typical null hypothesis states that there is no association between two variables in the population, that two or more samples are drawn from populations with the same mean or median, and so forth. In every case, the null hypothesis allows the statistician to designate a distribution of sample means or a similar probability distribution that can be used to assign probabilities to various outcomes of the research.

In our example, we developed our own distribution of sample means and were able to evaluate directly various outcomes of the experiment. For the statistical tests presented in this book, the probability distributions are represented by the tables of significance that accompany each test. All you need do is compute the value of the statistic and evaluate it against the appropriate tabled value. If the value of the test statistic is one that would occur very rarely (5% of the time or less) if the null hypothesis were true, you can reject the null hypothesis and conclude that there is a relationship between the variables you have studied. If you systematically adopt the 5% level, you will conclude that there is a difference when there is none only 5% of the time.

The steps in conducting a research study and evaluating the outcome with a statistical test are:

1. *Formulate your research hypothesis (ses).*

2. *Formulate your null hypothesis (ses) and alternative hypothesis (ses).*

3. *Design the experiment.* Select the independent variable(s) and dependent variable(s) of interest. Select the levels of the independent variable so that the impact on the dependent variable will be maximal. In general, *do not select too many levels.* Employ the largest feasible number of subjects.

4. *Select the appropriate statistical test.* The material presented in Chapters 2 and 3 and the branching program indicate the appropriate test for your research design.

5. *Gather the data.*

6. *Compute the value of your statistic.*

7. *Go to the table for your statistical test and find the value of the test statistic appropriate for your study.* Use an alpha level of .05 unless other *a priori* considerations indicate that this value is too high or too low. In general, use a two-tailed test.

8. *Accept or reject the null hypothesis based on the comparison of the obtained value with the tabled value for the level of alpha you have determined.*

9. *Interpret your findings in terms of your research hypothesis.*

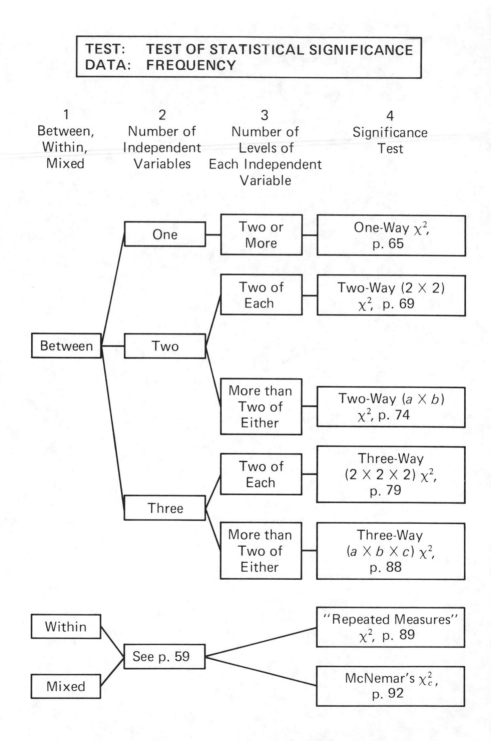

TEST: TEST OF STATISTICAL SIGNIFICANCE
DATA: FREQUENCY

1	2	3	4
Between, Within, Mixed	Number of Independent Variables	Number of Levels of Each Independent Variable	Significance Test

One — Two or More — One-Way χ^2, p. 65

Between

Two

Two of Each — Two-Way (2×2) χ^2, p. 69

More than Two of Either — Two-Way ($a \times b$) χ^2, p. 74

Three

Two of Each — Three-Way ($2 \times 2 \times 2$) χ^2, p. 79

More than Two of Either — Three-Way ($a \times b \times c$) χ^2, p. 88

Within

Mixed

See p. 59

"Repeated Measures" χ^2, p. 89

McNemar's χ_c^2, p. 92

FREQUENCY DATA

LIMITATIONS AND EXCEPTIONS

1. *Number of independent variables.* For frequency data, the dependent variable is the number of people in the various categories or cross-classifications of categories. In considering frequency data, it is simplest to regard all classification categories as independent variables.[1] We must be prepared to shift our focus slightly to permit our simplified conceptualization to fit all kinds of data. For example, if we conducted a study to determine whether differences in educational level were related to differences in income, we might classify a person's educational level and then measure his income. If we used numerical scores (score data) for income, educational level would be the independent variable and income the dependent variable. If, however, we classified people by predetermined income levels as well as by predetermined educational levels and counted how many people at each educational level occurred in each income grouping (frequency data), we would consider both educational level and income to be independent variables. Thus the same study is considered to have one independent variable if it uses score data but two independent variables if it uses frequency data.

2. If samples are very small, or if some expected events are extremely infrequent, the χ^2 tests presented on the following pages may not be appropriate. If the expected frequency in any cell is less than five, see the cautions on small samples on p. 62.

3. All χ^2 tests presented in this book require that each subject be counted only once (or, technically, that all frequencies be independent). If any subject has been counted more than once, see p. 60 for some suggestions on appropriate analyses.

---

[1] It is important to remember that this difficulty occurs only when the statistical technique needed is the significance test; it does not occur with correlation coefficients.

KINDS OF χ^2

All significance tests for frequency data in this book employ chi square (χ^2). Although other tests may be used with frequency data, χ^2 has the advantage of simplicity and has sufficient flexibility to adapt to a wide range of designs. Three basic χ^2 tests are illustrated in this chapter.

Test of association. The most frequent use of χ^2 is as a test of association. For this test, each subject is tallied in a cross-classification of two or more independent variables. For example, a subject may be classified as being a man or a woman and as having either blonde hair or brunette hair. If we randomly select 30 men and 30 women and then count the number of blondes and brunettes, forgetting about gender for the moment, we might find that 30 people were blondes and 30 were brunettes. If we construct a table (called a "contingency table") to present our gender-and-hair-color relationship, these values appear as marginal totals.

	Men	Women	
Blondes			30
Brunettes			30
	30	30	60

Let's assume that there is *absolutely no relationship between gender and hair color*. We can then estimate the way in which the subjects would be distributed within the cells of the table. There are 30 men, so the odds that any particular subject will be a man are 30/60 or 1/2. There are 30 brunettes, so the odds that any particular subject will be a brunette are 30/60 or 1/2. If gender and hair color are completely independent, the probability that an individual will be both male and brunette equals the product of the two independent probabilities, or $1/2 \times 1/2 = 1/4$. Since there are 60 people, 1/4 of them, or 15, would be brunette men if hair color and gender are independent. Similarly, we would expect 15 blonde men, 15 blonde women, and 15 brunette women. These values are the *expected frequencies* (*fe*) if there is no association between hair color and gender. The null hypothesis of no association between gender and hair color is being tested and the equality of the expected frequencies is merely a quantitative statement of that null hypothesis.

In the most clear-cut association between the two variables, all men are brunettes and all women are blondes, or vice versa. In that case, there would be maximum deviation of the observed frequencies (*fo*) from the expected frequencies in each cell. The more the observed frequencies deviate from the expected frequencies, the less likely it is that the null hypothesis is true. All two-way χ^2, three-way χ^2, and repeated-measures χ^2 provide tests of association.

Deviation from a priori *expected frequencies*. In some cases, the researcher may not be interested in an association between two variables. Instead, you may wish to determine whether your observed frequencies deviate significantly from some prior expectations. χ^2 may be used to make this determination. For example, a public-opinion poll may show that 275 respondents say they will vote for Jones for city council and 225 say they will vote for Smith. You may wonder whether these proportions represent a significant deviation from a 50-50 split. In that case, you would assign an expected frequency of 250 votes to each man and do a χ^2 to determine whether the observed frequencies differ significantly from the expected. Or you may know that, in the last election, Jones got 60% of the vote and Smith received 40%. To determine whether the observed frequencies represent a significant deviation from a 60-40 split, you would assign an expected frequency of 300 to Jones and an expected frequency of 200 to Smith.

In either case, the assignment of expected frequencies represents a quantitative statement of the null hypothesis. In the first case, the null hypothesis is that the subjects come from a population in which the votes are split 50-50. In the second case, the null hypothesis is that the subjects come from a population in which the votes are split 60% for Jones and 40% for Smith.

Although the researcher usually designs a study specifically to test deviation from *a priori* expected frequencies, questions about such deviations may also be asked in conjunction with the test of association. Following a χ^2 test of association, you may sometimes wish to determine whether the *marginal totals* of the contingency table deviate from some expected values. For instance, (going back to our earlier example), you may wish to ask whether there were an equal number of blondes and brunettes. This question is quite appropriate. In testing marginal totals, as well as in the usual test of *a priori* expected frequencies, you may test any null hypothesis that interests you. The one-way χ^2 provides a test of deviation from *a priori* expected frequencies in both of these cases.

Repeated measures (*within-subjects*) χ^2. Although all scores must be independent to permit the use of χ^2, we have included this section on repeated-measures χ^2 because there are a number of situations that resemble repeated measures so closely that they are likely to confuse beginners. We discuss two basic χ^2 designs that on superficial examination appear to involve repeated measures of a single subject or event. The first provides a test of association; the second provides a measure of change of the proportion of subjects in the conditions between the measurements. However, it should be emphasized that, if subjects have been counted two or more times, and your analysis is not analogous to one of those discussed below, a χ^2 analysis is not appropriate. This subject is discussed in more detail in a later section.

In the first case, at Time 1, an individual is classified into either category A_1 or category A_2. At Time 2, he is classified again into either A_1 or A_2. A χ^2 analysis is appropriate if the data are arranged as a 2 X 2 matrix with each individual counted (once) in A_1A_1, A_1A_2, A_2A_1, or A_2A_2. For example, suppose we get a random sample of eligible voters and ask, after the presidential election (Time 1), whether they voted. Two years later, following the off-presidential-year elections (Time 2), we ask the same voters whether they voted. Each time, we classify the eligible voters as voters or nonvoters. Their behavior may then be cross-classified as voter-voter, voter-nonvoter, nonvoter-voter, or nonvoter-nonvoter. Thus each subject is cross-classified on the basis of his classification at Time 1 and at Time 2. This cross-classification is identical to that encountered in any test-of-association χ^2. An example and computing rules for this type of analysis are given in the section for *Repeated-Measures Chi Square: Test of Association*, later in this chapter.

Any design that involves three repeated measures, or two repeated measures and one between-subjects measure (mixed design), may be analyzed using the three-way χ^2, *provided that all frequencies are independent in the cross-classification*. To expand the example given above, we might examine the behavior of eligible voters who differ on some other characteristic, such as gender. Gender would be a between-subjects factor. We thus have a 2 X 2 X 2 χ^2 with voting behavior at Time 1, voting behavior at Time 2 (cross-classified for each subject), and gender of subject. No worked example or computing rules are given for these infrequent designs, but you can determine the procedures by studying the repeated-measures χ^2 to determine how to set up the analysis and then following exactly the computing rules given for the three-way χ^2 in the analysis later in this chapter.

An alternative question that may be asked when each subject is measured twice on the same dimension[2] is whether the proportion

---

[2] The two observations may involve matched pairs, litter mates, twins, or two observations on the same individual. That is, this test does not require the two samples to be independent. For simplicity, we consider the test only as it applies to two observations on the same individual, but an analogous interpretation is possible for the other cases.

of people in Condition 1 on the first measurement is different from the proportion of people in Condition 1 on the second measurement. McNemar's χ_c^2 tests the null hypothesis that the proportion in Condition 1 on the first occasion equals the proportion in Condition 1 on the second occasion. For example, we might ask whether the proportion of people who voted in one election was the same as the proportion who voted in a second election.

An example and computing rules for McNemar's χ_c^2 are given on p. 92. If the measures are repeated more than once, the Cochran test is appropriate.[3]

RESTRICTIONS ON THE USE OF χ^2

There are a number of restrictions on the use of χ^2. Violation of the first two restrictions is the source of the most common mistakes made by beginning researchers.

The raw data for χ^2 must always be frequencies. χ^2 is appropriate only when you count the number of subjects in particular classifications or cross-classifications. It is perfectly appropriate, for example, to apply χ^2 to a study in which you count the number of boys and girls who pass or fail a particular test. Other times you may count the number of items that each boy or girl passes and get an average score for boys and girls. A χ^2 analysis may *not* be applied to these data. Unfortunately, the means for two such groups may look enough like frequencies that a novice will perform a χ^2 analysis on the data.[4] Since in the latter case each person has a score, an analysis designed for score data must be employed. If data are presented as percentages or proportions of people who fall into categories, they may be easily converted to frequencies by multiplying the total number of subjects by the percentages or proportions. χ^2 analysis may then be performed.

All χ^2 analyses require that each subject or event be counted only once (or, technically, that all frequencies be independent). If we wish to find out whether boys or girls are more likely to pass or fail a test, we might observe the performance of 100 children on a test. We may *not* observe the performance of 25 children on four tests, however, because the four tests for each child may be expected to be related.

This requirement would appear to rule out completely the use of χ^2 for within-subjects (repeated-measures) designs, since each subject would be counted more than once. However, *even if people are measured two or more times, χ^2 may be used if the design can be*

---

[3] An explanation of the Cochran test is presented in Wike (1971), pp. 135–136.

[4] Spence, et al. (1968), p. 199, provides a simple explanation of why an analysis on means will not work.

arranged so that each subject is counted only once in the contingency table (and thus the assumption of independence is not violated). Designs in which subjects are measured more than once but the measurements are still independent were discussed in the section *Repeated-Measures (Within-Subjects)* χ^2.

If samples are very small, or if some expected events are extremely infrequent, χ^2 *may not be appropriate.* For contingency tables with 1 *df* (one-way with two levels, or 2 X 2), all expected frequencies must be five or more. For contingency tables having more than 1 *df* (2 X 3 or larger), all expected frequencies must be two or greater. Rules for computing expected frequencies are included with the computational procedures.

If any expected frequency is less than five for χ^2 with 1 *df*, more data may be gathered or Fisher's Exact Test may be used. Fisher's Exact Test requires either tedious computations or extended tables (and no calculations). [5] If any expected frequency is less than two for χ^2 with more than 1 *df*, Fisher's Exact Test is not appropriate. In this case, more data may be gathered or the number of categories may be collapsed (if there is some logical basis for this procedure) to produce categories in which all expected frequencies are greater than two.

Categories should be set up before the experiment is run on the basis of some logical classification. Most common categories, such as gender, age, income, or occurrence or nonoccurrence of events, are perfectly legitimate. The decision to develop categories after the study has been performed in order to take advantage of relationships that appear to exist in the data is illegitimate. The problems with *post hoc* tests are discussed in Chapter 10.

Whenever the frequency of the occurrence of an event is recorded, it is essential that the frequency of the nonoccurrence of the event be recorded as well. Both frequencies are required for computation of χ^2. For example, if you wish to study the relationship of gender to passing or failing a test, you must record in your sample not only those who passed but also those who failed. This requirement is a restatement of the requirement that the expected and obtained frequencies be equal.

---

[5] See Wike (1971), pp. 126–127, for a full discussion of Fisher's Exact Test and appropriate tables.

NOTES ON DEGREES OF FREEDOM

A final word about degrees of freedom (df) for χ^2 must be added. A value of χ^2 cannot be evaluated unless the number of degrees of freedom associated with it is known.[6] The number of degrees of freedom associated with any χ^2 may be easily computed as follows:

If there is one independent variable: $df = a - 1$ where a is the number of levels of the independent variable.

If there are two independent variables: $df = (a - 1)(b - 1)$ where a and b are the number of levels of the first and second independent variables, respectively.

If there are three independent variables: $df = (a - 1)(b - 1)(c - 1)$ where a, b, and c are the number of levels of the first, second, and third independent variables, respectively.

There is, in addition, a very simple way of thinking about degrees of freedom that does not involve computing formulas. The degrees of freedom for χ^2 indicate the number of cells that are free to vary after the marginal totals have been fixed. In our example on the association between gender and hair color, we found that the marginal total for men was 30, for women was 30, for blondes was 30, and for brunettes was 30. If we are told that there are 4 blonde women in the sample of 60, we may immediately determine the frequencies in the other three cells. Since there are 30 women in all and 4 are blondes, 26 must be brunettes. Since there are 30 blondes in all and 4 are women, 26 must be men. Finally, since there are 30 men in all and 26 are blondes, 4 must be brunettes. Only one cell is free to vary in a 2 × 2 table because as soon as it is fixed, all other cell values are known. Since the number of cells free to vary determines the df, a 2 × 2 table always has only one df.

The same reasoning applies to degrees of freedom for any other χ^2. If the marginal totals in a 3 × 3 table are fixed, four cells can vary before all of the other cell values are known. In a one-way table with four levels, if the total is fixed, three cells may vary before the final cell is fixed. Degrees of freedom for these two tables will be four and three respectively.

SUMMARY

This brief introduction should demonstrate that the χ^2 test is quite simple. The researcher starts with a set of observed frequencies.

---

[6] The shape of the underlying sample distribution of χ^2, against which values are evaluated, changes as the number of degrees of freedom increases.

Expected frequencies are determined either by some *a priori* expectations or on the assumption that the independent variables are not related. These expected frequencies represent the null hypothesis being tested. The more the observed frequencies differ from the expected frequencies, the less likely it is that the null hypothesis is true.

NOTE ON THE ANALYSES AND WORKED EXAMPLES

Each analysis is presented with spelled-out step-by-step procedures. There is a worked example for most analyses. The example (1) begins with a description of the study, (2) presents an appropriate null hypothesis, (3) indicates the steps through the branching program, (4) goes step by step through the computations, and (5) presents the statistical and research outcomes of the study.

The three-way χ^2 with a levels is not presented in complete detail because of its length. In addition, because of the length and complexity of the computations, no example is provided for this three-way χ^2 test.

Since later computations refer back to steps by number (for example, "#4 less #5"), you will save time if you list the products from each step by number as you proceed. In addition, you can locate and correct errors in your computations more easily if you have recorded your intermediate steps systematically.

Accompanying each analysis is a brief description that lists in summary form when the analysis may be used, special requirements, and so on. Each analysis provides a reference to the chapters on interpretation, specific comparisons, and strength of association. Each example includes both a statistical conclusion and a journal-style interpretation. We have not included either the specific-comparison tests or the strength-of-association measures for these examples.

Description:

Deviation from
a priori fe
One independent
variable
Two levels: use
correction for
contingency
a levels: use regular
formula
Between-subjects
Specific compari-
sons available
Strength of associ-
ation not
appropriate

ONE-WAY CHI SQUARE

ONE INDEPENDENT VARIABLE

WITH TWO OR MORE LEVELS

Follow through the steps below.

- *Have you read the* Limitations and Exceptions *section?*

- *Are all of these assumptions met for your study? If not, do not continue.*

- *This analysis is not appropriate when* df $= 1$ *if any* fes *are less than five, or when* df $= 2$ *if any* fes *are less than two.*

- *If you are experienced with computations for* χ^2, *go directly to the computing formula below and proceed. If you require step-by-step guidance, begin with Section B,* The Example.

A. COMPUTING FORMULA

1. *Three or More Levels of One Independent Variable*

$$\chi^2 = \sum_{A=1}^{a} \frac{(fo - fe)^2}{fe}$$

χ^2 table, Appendix 1
$df = a - 1$

where $a =$ the number of levels of the independent variable
(the number of cells),

$fo =$ the observed frequency for each of the *a* cells, and

$fe =$ the expected frequency for each of the *a* cells,
based on some theoretical expectation (H_0).

2. *Two Levels of the Independent Variable*

IMPORTANT: If there are only two levels of the independent variable, a correction for continuity must be used. The computing formula then becomes

$$\chi^2 = \sum_{A=1}^{a} \frac{(|fo - fe| - .5)^2}{fe} \qquad \begin{array}{l} \chi^2 \text{ table, Appendix 1} \\ df = 1 \end{array}$$

B. THE EXAMPLE

1. A public-opinion polling team in a small town was interested in the type of sporting events that adults in the age bracket 20-50 years prefer to watch on TV. A random sample of 120 in this age bracket was selected and asked, "Given your preference, would you prefer to watch baseball, basketball, or football on your television set at home?" Of the respondents, 39 indicated a preference for baseball, 25 selected basketball, and 56 selected football.

2. *Null Hypothesis*
   a. *General.* In the population being sampled, the proportions of people in each category are equal. $H_0 : p_1 = p_2 = p_3$.

   NOTE: This is only one of an infinite number of null hypotheses that may be tested. The hypothesis of equality of the cells is the most common.

   b. *Specific.* In the population being sampled, equal proportions of people prefer baseball, basketball, and football.

3. *Steps through the Branching Program for This Example*
   a. *Statistical technique needed*: **test of significance**
   b. *Kind of data*: **frequency**. The number of people in each category is counted.
   c. *Between, within, mixed*: **between**. Each person is tallied in the category of his or her favorite sport; therefore each person is measured once.
   d. *Number of independent variables*: **one**. Category of sport preferred
   e. *Number of levels of the independent variable*: **three**. Baseball, basketball, football

C. RECORDING THE DATA

Record the data in a table with *a* columns. Our example comprises a single classification study with three levels of the independent variable.

Preferred Sport

	Baseball	Basketball	Football	Total
Observed Frequencies	39	25	56	120
Expected Frequencies	40	40	40	

D. COMPUTATIONS

1. Select appropriate *fes*. The researcher must select expected frequencies for each cell. These are based on the researcher's expectations prior to the study.

> The polling team expected the three sports to be chosen equally often. Thus $fe_{A_1} = 40$, $fe_{A_2} = 40$, $fe_{A_3} = 40$. In other words, the expected frequency for football was 40, for baseball was 40, and for basketball was 40. (If the polling team had expected football to be chosen twice as often as either of the other sports, the expected frequencies would be football, 60; baseball, 30; and basketball, 30.)

CHECK: $\Sigma fe = \Sigma fo = N$

Sum the expected frequencies; sum the observed frequencies. Both sums must equal N, the number of subjects.

> $40 + 40 + 40 = 39 + 25 + 56 = 120$ OK!

CHECK: Are all *fes* greater than five for $df = 1$, or greater than two for $df = 2$ or more?

> $40 > 2$ OK!

2. Select appropriate computing formula and solve for χ^2:
   a. *If the number of levels of the independent variable is three or more:*

$$\chi^2 = \frac{(fo_{A_1} - fe_{A_1})^2}{fe_{A_1}} + \frac{(fo_{A_2} - fe_{A_2})^2}{fe_{A_2}} + \frac{(fo_{A_3} - fe_{A_3})^2}{fe_{A_3}}$$

For each cell: Take the observed frequency, subtract the expected frequency, square the outcome, and divide by the expected frequency. Then sum the terms from each cell. (If *fes* are equal, the squared outcomes may be summed and then divided by *fe*, as in the example.)

$$\frac{(39-40)^2}{40} + \frac{(25-40)^2}{40} + \frac{(56-40)^2}{40} = \frac{(-1)^2}{40} + \frac{(-15)^2}{40} +$$

$$\frac{(16)^2}{40} = \frac{1}{40} + \frac{225}{40} + \frac{256}{40} = \frac{482}{40} = 12.05$$

b. *If the number of levels of the independent variable is two:*

$$\chi^2 = \frac{(|fo_{A_1} - fe_{A_1}| - .5)^2}{fe_{A_1}} + \frac{(|fo_{A_2} - fe_{A_2}| - .5)^2}{fe_{A_2}}$$

For each cell: Take the observed frequency, subtract the expected frequency, drop the plus or minus sign, and *reduce this absolute value by .5*; square the outcome, divide by the expected frequency, then sum the terms from both cells.

CHECK: $\Sigma(fo - fe) = 0$
The sum of the observed frequencies minus the expected frequencies must equal 0.

$$(39-40) + (25-40) + (56-40) = (-1) + (-15) + (16) = 0 \qquad \text{OK!}$$

3. Enter the χ^2 table, Appendix 1, with $df = a - 1$. Note the tabled value of χ^2 for $\alpha = .05$.

$df = a - 1 = 3 - 1 = 2$. Tabled value of χ^2 with 2 *df* and
$\alpha = .05$ is 5.99.

4. a. If $\chi^2_{\text{Computed}} \geq \chi^2_{\text{Tabled}}$ (if #2 is equal to or larger than #3), then the deviation of the observed frequencies from the expected frequencies is significant. Reject the null hypothesis.
 b. If $\chi^2_{\text{Computed}} < \chi^2_{\text{Tabled}}$ (if #2 is smaller than #3), then the deviation of the observed frequencies from the expected frequencies is not significant. Do not reject the null hypothesis.

12.05 > 5.99. Therefore the deviation of the observed frequencies from the expected frequencies is significant.
Reject the null hypothesis.

E. INTERPRETATION OF THE EXAMPLE

1. *Statistical conclusion.* If these observations were drawn randomly from a population in which the three sports were equally popular, cell frequencies as discrepant as these would occur less than 5% of the time by chance. Therefore the null hypothesis should be rejected.
2. *Researcher's conclusion.* The three sports are not equally popular for the population. (Football is chosen more often and basketball is chosen less often than would be expected by chance if the three sports were equally popular.)

F. FURTHER PROCEDURES

1. Interpretation: see Chapter 9, *Introduction* and Section A.
2. Specific comparisons: see Chapter 10, *Introduction* and Section A.
3. Strength-of-association measure: not appropriate.[7]

Description:

Test of association
Two independent
  variables
Two levels of each
  independent
  variable
Between-subjects
Simplified version
  of $a \times b \; \chi^2$
Specific comparisons:
  not necessary
Strength of associa-
  tion available

TWO-WAY CHI SQUARE ($2 \times 2 \; \chi^2$)

TWO INDEPENDENT VARIABLES

WITH TWO LEVELS OF EACH

---

[7]Strength of association does not make sense in this case. What is the association between? It would have to be between observed and expected frequencies, which is a meaningless association.

Follow through the steps below.

● *Have you read the* Limitations and Exceptions *section?*

● *Are all of these assumptions met for your study? If not, do not continue.*

● *This analysis is not appropriate if any* fes *are less than five. See p. 62.*

● *If you are experienced with computations for* χ^2, *go directly to the computing formula below and proceed. If you require step-by-step guidance, begin with Section B,* The Example.

A. COMPUTING FORMULA, TWO INDEPENDENT VARIABLES WITH TWO LEVELS OF EACH

$$\chi^2 = \frac{N(|bc - ad| - N/2)^2}{(a + b)(c + d)(a + c)(b + d)}$$

χ^2 table, Appendix 1
$df = 1$

where $a, b, c, d,$ = the observed frequencies for each of the four cells of the 2 × 2 matrix (as shown below), and

N = the number of cases.

A

	A_1	A_2	
B_1	a	b	$a + b$
B_2	c	d	$c + d$
	$a + c$	$b + d$	$N = a + b + c + d$

(*B* labels the rows B_1 and B_2.)

NOTE: No computation of expected frequencies is necessary when using this computing formula unless you suspect that any fe is less than 5. Expected frequencies are computed by multiplying the row total by the column total and dividing by N for each cell. They should be computed if any expected frequency might be less than five.

B. THE EXAMPLE

1. A psychologist predicted gender differences in solving a reasoning problem. She randomly selected 36 men and 39 women and administered the problem to them. Of the 36 men, only 10 succeeded in solving the problem. Of the 39 women, 26 succeeded.

2. *Null Hypothesis*
   a. *General.* In the population being sampled, the rows and columns are independent. For example, $H_0 : p(A_1/B_1) = p(A_1/B_2)$.
   b. *Specific.* In the population being sampled, the same proportion of men and women are able to solve the reasoning problem.
3. *Steps through the Branching Program for This Example*
   a. *Statistical technique needed:* **test of significance**
   b. *Kind of data:* **frequency**. The number of people in each category is counted.
   c. *Between, within, mixed:* **between**. Each person is either male or female and may either succeed or fail.
   d. *Number of independent variables:* **two**. Gender and success
   e. *Number of levels of each independent variable:* **gender has two levels: male and female. Success has two levels: success or failure.**

C. RECORDING THE DATA

Record the data in a 2 × 2 matrix in which A designates one independent variable and B designates the other. Starting in the upper left-hand cell, label each of the four cells a, b, c, and d. The total observed frequency for a particular row or column is the marginal frequency for that row or column. The marginal frequencies are designated $a + b$, $c + d$, $a + c$, and $b + d$.

	Fail	*Succeed*	
Women	13 (a)	26 (b)	39
Men	26 (c)	10 (d)	36
	39	36	75

D. COMPUTATIONS

1. $|bc - ad|$
   Multiply the observed frequency of b by that of c. From this, subtract the product of the observed frequencies of cells a and d. Drop the plus or minus sign.

$$|(26 \times 26) - (13 \times 10)| = |676 - 130| = 546$$

2. $|bc - ad| - N/2$

From this absolute value subtract $N/2$, where N is the total number of cases.

$$546 - 75/2 = 546 - 37.5 = 508.5$$

3. $(|bc - ad| - N/2)^2$. Square #2.

$$(508.5)^2 = 258,572.25$$

4. $N(|bc - ad| - N/2)^2$. Multiply #3 by N.

$$(75)(258,572.25) = 19,392,918.75$$

5. $(a + b)(c + d)(a + c)(b + d)$

Obtain the product of the four marginal frequencies.

$$(39)\,(36)\,(39)\,(36) = 1,971,216$$

6. $\chi^2 = \dfrac{N(|bc - ad| - N/2)^2}{(a + b)(c + d)(a + c)(b + d)}$. Divide #4 by #5.

$$\frac{19,392,918.75}{1,971,216} = 9.84$$

CHECK: Are all *fe*s greater than five?

The cell with the smallest *fe* associated with it will be the cell with the smallest row and column totals. It is evident from inspection that d will have the smallest *fe*. Although it does not look suspiciously small, we have computed it. If it is larger than five, all other cells will also be larger than five.

$$fe_d = \frac{(36)(36)}{75} = 17.2 > 5. \quad \text{OK!}$$

CHECK: $a + b + c + d = N$

$$13 + 26 + 26 + 10 = 75$$

7. Tabled value of χ^2 with 1 *df* and $\alpha = .05$ is 3.84.
8. a. If $\chi^2_{\text{Computed}} \geq \chi^2_{\text{Tabled}}$ (if #6 is equal to or larger than #7), then there is a significant association between the two independent variables. Reject the null hypothesis.
   b. If $\chi^2_{\text{Computed}} < \chi^2_{\text{Tabled}}$ (if #6 is smaller than #7), then the association between the two independent variables is not significant. Do not reject the null hypothesis.

9.84 > 3.84. Therefore there is an association between gender of the subjects and frequency of solving the problem.
Reject the null hypothesis.

E. INTERPRETATION OF THE EXAMPLE

1. *Statistical conclusion.* If these observations were drawn randomly from a population in which the proportions of men and women who solved the problem were equal, cell frequencies as discrepant as these would occur less than 5% of the time by chance. Therefore the null hypothesis should be rejected.
2. *Researcher's conclusion.* The proportion of women who are able to solve this problem is significantly greater than the proportion of men who are able to solve it.

F. FURTHER PROCEDURES

1. Interpretation: see Chapter 9, *Introduction* and Section A.
2. Specific comparisons: in some cases, the researcher may wish to test the marginal frequencies against some theoretical expectations (see p. 59). No other specific comparisons are necessary.
3. Strength-of-association measure: see Chapter 11, *Introduction* and Section A.

Description:

TWO-WAY CHI SQUARE ($a \times b \chi^2$)
TWO INDEPENDENT VARIABLES
WITH MORE THAN TWO LEVELS
OF EITHER VARIABLE

Test of association
Two independent
   variables
More than two levels
   of either variable
Between-subjects
Specific comparisons
   and strength of
   association
   available

Follow through the steps below.

- *Have you read the* Limitations and Exceptions *section?*

- *Are all of these assumptions met for your study? If not, do not continue.*

- *This analysis is not appropriate if any* fes *are less than two. See p. 62.*

- *If you are experienced with computations for* χ^2, *go directly to the computing formula below and proceed. If you require step-by-step guidance, begin with Section B,* The Example.

A. COMPUTING FORMULA, GENERAL FORM χ^2 FOR TWO INDEPENDENT VARIABLES WITH MORE THAN TWO LEVELS OF EITHER OR BOTH

$$\chi^2 = \sum_{AB=1}^{ab} \frac{(fo - fe)^2}{fe} \qquad \begin{array}{l} \chi^2 \text{ table, Appendix 1} \\ df = (a-1)(b-1) \end{array}$$

where a = the number of levels of independent variable A,
   b = the number of levels of independent variable B,
   ab = the total number of cells,
   fo = the observed frequency for each of the ab cells, and
   fe = the expected frequency for each of the ab cells.

B. THE EXAMPLE

1. The student government of a small university conducted a referendum to determine whether or not the students were willing to pay higher activities fees to increase the number of cultural events held on campus. The referendum results were very close. The editor of the school newspaper decided to take a poll

to see whether there was any relationship between the students' living accommodations (fraternity house, dormitory, commuter) and their position on the issue. He randomly selected 40 commuters, 30 dorm dwellers, and 30 fraternity members and asked them whether they approved, disapproved, or had no opinion about the proposed fee change.

2. *Null Hypothesis*

   a. *General.* In the population being sampled, the rows and columns are independent.

   b. *Specific.* In the population being sampled, the proportion approving, disapproving, or having no opinion on the fee change is the same among fraternity members, dorm dwellers, and commuters.

3. *Steps through the Branching Program for This Example*

   a. *Statistical technique needed:* **test of significance**

   b. *Kind of data:* **frequency.** For each living classification, count the number who approve, disapprove, or have no opinion on the fee increase.

   c. *Between, within, mixed:* **between.** Each person has one living classification and has only one opinion on the fee increase; thus each person is tallied only once in the table.

   d. *Number of independent variables:* **two.** Living classification and attitude on the fee increase

   e. *Number of levels of each independent variable:* **living classification has three levels: fraternity house, dormitory, commuter. Attitude has three levels: approve, disapprove, no opinion.**

C. RECORDING THE DATA

Record the data in an $a \times b$ table. Our example comprises a two-way χ^2 with three levels of each independent variable (3 × 3 table).

	Opinion on Issue			
	Approve (A_1)	Disapprove (A_2)	No Opinion (A_3)	Marginal Totals
Commuters (B_1)	5	20	15	40
Dorm Dwellers (B_2)	20	5	5	30
Fraternities (B_3)	10	10	10	30
Marginal Totals	35	35	30	100

Living Accommodations

D. COMPUTATIONS

1. Compute appropriate *fe*s.

$$fe_{A_1 B_1} = \frac{(fo_{A_1})(fo_{B_1})}{N}, \qquad fe_{A_1 B_2} = \frac{(fo_{A_1})(fo_{B_2})}{N}, \qquad \text{and so on.}$$

For each cell, the expected frequency is computed by multiplying the row total for that cell by the column total for that cell and dividing by N, the total number of people.[8] This operation must be performed for each cell using the appropriate row and column totals in each computation. (There will be nine such computations in a $3 \times 3 \chi^2$.)

$$fe_{A_1 B_1} = \frac{(35)(40)}{100} = \frac{1400}{100} = 14 \qquad fe_{A_2 B_3} = \frac{(35)(30)}{100} = \frac{1050}{100} = 10.5$$

$$fe_{A_1 B_2} = \frac{(35)(30)}{100} = \frac{1050}{100} = 10.5 \qquad fe_{A_3 B_1} = \frac{(30)(40)}{100} = \frac{1200}{100} = 12$$

$$fe_{A_1 B_3} = \frac{(35)(30)}{100} = \frac{1050}{100} = 10.5 \qquad fe_{A_3 B_2} = \frac{(30)(30)}{100} = \frac{900}{100} = 9$$

$$fe_{A_2 B_1} = \frac{(35)(40)}{100} = \frac{1400}{100} = 14 \qquad fe_{A_3 B_3} = \frac{(30)(30)}{100} = \frac{900}{100} = 9$$

$$fe_{A_2 B_2} = \frac{(35)(30)}{100} = \frac{1050}{100} = 10.5$$

CHECK: $\Sigma fe = \Sigma fo = N$

The sum of the expected frequencies must equal the sum of the obtained frequencies, which must equal the total number of cases.

$$14 + 10.5 + 10.5 + 14 + 10.5 + 10.5 + 12 + 9 + 9 = 5$$
$$+ 20 + 10 + 20 + 5 + 10 + 15 + 5 + 10 = 100 \qquad \text{OK!}$$

CHECK: Are all *fe*s greater than two? OK!

[8] Conceptually, the procedure involves multiplying the probability of being in a given row by the probability of being in a given column and multiplying that product by N. Thus for cell $A_1 B_1$ the *fe* is $(fo_{A_1}/N)(fo_{B_1}/N)(N)$. This equation simplifies to the equation $[(fo_{A_1})(fo_{B_1})]/N$.

2. Solve for χ^2 :

$$\chi^2 = \frac{(fo_{A_1B_2} - fe_{A_1B_2})^2}{fe_{A_1B_2}} + \frac{(fo_{A_1B_2} - fe_{A_1B_2})^2}{fe_{A_1B_2}} + \frac{(fo_{A_1B_3} - fe_{A_1B_3})^2}{fe_{A_1B_3}}$$

$$+ \cdots + \frac{(fo_{A_2B_2} - fe_{A_2B_2})^2}{fe_{A_2B_2}} + \cdots + \frac{(fo_{A_3B_3} - fe_{A_3B_3})^2}{fe_{A_3B_3}}$$

For each cell: take the observed frequency, subtract the expected frequency, square the outcome, and divide it by the expected frequency. Then obtain a total by summing the terms for all cells. (In a 3 X 3 table, there will be nine terms in the equation, one for each cell.)

$$\frac{(5 - 14)^2}{14} + \frac{(20 - 10.5)^2}{10.5} + \frac{(10 - 10.5)^2}{10.5}$$

$$+ \frac{(20 - 14)^2}{14} + \frac{(5 - 10.5)^2}{10.5} + \frac{(10 - 10.5)^2}{10.5}$$

$$+ \frac{(15 - 12)^2}{12} + \frac{(5 - 9)^2}{9} + \frac{(10 - 9)^2}{9}$$

$$= \frac{81}{14} + \frac{90.25}{10.5} + \frac{.25}{10.5} + \frac{36}{14} + \frac{30.25}{10.5} + \frac{.25}{10.5} + \frac{9}{12} + \frac{16}{9} + \frac{1}{9}$$

$$= \frac{117}{14} + \frac{9}{12} + \frac{121}{10.5} + \frac{17}{9}$$

$$= 8.36 + .75 + 11.53 + 1.89 = 22.53$$

CHECK: $\Sigma(fo - fe) = 0$

The sum of the observed frequencies minus the expected frequencies must equal 0.

$$(5 - 14) + (20 - 10.5) + (10 - 10.5) + (20 - 14) + (5 - 10.5)$$
$$+ (10 - 10.5) + (15 - 12) + (5 - 9) + (10 - 9) = 0$$

3. Enter the χ^2 table, Appendix 1, with $df = (a - 1)(b - 1)$. Note the tabled value of χ^2 for $\alpha = .05$.

$df = (3 - 1)(3 - 1) = 4$. The tabled value of χ^2 with 4 *df* and
$\alpha = .05$ is 9.488.

4. a. If $\chi^2_{\text{Computed}} \geq \chi^2_{\text{Tabled}}$ (if #2 is equal to or larger than #3),
      then there is a significant association between the two inde-
      pendent variables. Reject the null hypothesis.
   b. If $\chi^2_{\text{Computed}} < \chi^2_{\text{Tabled}}$ (if #2 is smaller than #3), then the
      association between the two independent variables is not
      significant. Do not reject the null hypothesis.

22.53 > 9.488. Therefore the deviation of the observed frequencies from the expected frequencies is significant. Reject
the null hypothesis.

E. INTERPRETATION OF THE EXAMPLE

1. *Statistical conclusion.* If these observations were drawn randomly from a population in which the proportions approving,
   disapproving, or having no opinion on the fee change were the
   same among fraternity members, dorm dwellers, and commuters,
   cell frequencies as discrepant as these would occur less than 5%
   of the time by chance. Therefore the null hypothesis should be
   rejected.
2. *Researcher's conclusion.* Among the student body at the university, opinion on the issue of raising activities fees is related
   to the students' living accommodations. (If you wish to make
   more specific statements about living accommodations and
   associated attitudes, specific-comparison tests must be
   performed.)

F. FURTHER PROCEDURES

1. Interpretation: see Chapter 9, *Introduction* and Section A.
2. Specific comparisons: in some cases, the researcher may wish
   to test the marginal frequencies against some theoretical expectations (see p. 59.) See also Chapter 10, *Introduction* and
   Section A.
3. Strength-of-association measure: see Chapter 11, *Introduction*
   and Section A.

Description:

Test of association
Three indepen-
   dent variables
Two levels of each
   variable
Between-subjects
Specific comparison:
   not necessary
Strength of association
   available

THREE-WAY CHI SQUARE ($2 \times 2 \times 2\chi^2$)

THREE INDEPENDENT VARIABLES

WITH TWO LEVELS OF EACH

Follow through the steps below.

- *Have you read the* Limitations and Exceptions *section?*

- *Are all of these assumptions met for your study? If not, do not continue.*

- *This analysis is not appropriate if any* fes *are less than five. See p. 62.*

- *If you are experienced with computations for* χ^2, *go directly to the computing formulas below and proceed. If you require step-by-step guidance, begin with Section B*, The Example.

A. COMPUTING FORMULAS, THREE-WAY χ^2 WITH TWO LEVELS OF EACH OF THE THREE INDEPENDENT VARIABLES

1. Compute one χ^2 for each possible combination of any two independent variables taken together. If the three independent vari-

ables are designated A, B, and C, you will then have one χ^2 reflecting the $A \times B$ comparison, one χ^2 reflecting the $A \times C$ comparison, and one χ^2 reflecting the $B \times C$ comparison. Use the following formula for these computations.

NOTE: No correction for continuity is used in three-dimensional χ^2s.

$$\chi^2 = \frac{N(bc - ad)^2}{(a + b)(c + d)(a + c)(b + d)}$$
χ^2 table, Appendix 1
$df = 1$.

where a, b, c, and d = the observed frequencies of each of the four cells generated by a particular 2×2 comparison, and

N = the total number of cases in a particular comparison.

2. Compute χ^2_{Total} using the following formula:

$$\chi^2_{Total} = \sum_{ABC=1}^{abc} \frac{(fo - fe)^2}{fe}$$

where a, b, and c = the number of levels of independent variables A, B, and C, respectively,

abc = the total number of cells,

fo = the observed frequency of each of the abc cells, and

fe = the expected frequency of each of the abc cells.

3. Compute one χ^2 reflecting a simultaneous comparison among all three independent variables ($A \times B \times C$). Use the following formula:

$$\chi^2_{A \times B \times C} = \chi^2_{Total} - \chi^2_{A \times B} - \chi^2_{A \times C} - \chi^2_{B \times C}$$
χ^2 table, Appendix 1
$df = 1$

B. THE EXAMPLE

1. A breeder of exotic tropical fish noticed that his male guppies differed from one another along three main dimensions. Some had a black border around their tails; others did not. Some had a red chin; others had a colorless chin. Finally some had blue bars on their sides; others did not. He wondered whether these characteristics were associated with one another. After the next breeding period, he selected at random 75 young male guppies and classified them on these three dimensions.

2. *Null Hypothesis*
   a. *General.* There is no association for the three characteristics in the population being sampled.
   b. *Specific.* Any fish may have a black border, a red chin, or a blue bar; however, in the population being sampled, any of these characteristics is independent of the other two (considered either singly or together).
3. *Steps through the Branching Program for This Example*
   a. *Statistical technique needed:* **test of significance**
   b. *Kind of data:* **frequency.** Guppies are classified on three dimensions and the number in each category is counted.
   c. *Between, within, mixed:* **between.** Each fish is counted only once in the subclassification that describes him: for example, black border, red chin, blue bars. He can occur in no other category.
   d. *Number of independent variables:* **three.** Color of border, color of chin, color of bars.
   e. *Number of levels of each independent variable:* **color of border has two levels: black and not black. Color of chin has two levels: red and not red. Color of bars has two levels: blue and not blue.**

C. RECORDING THE DATA

1. Construct a table giving all eight possible $A \times B \times C$ combinations (see Table 5.1). For each combination, record the corresponding observed frequency in Column *fo*. In Column *fe*, record the corresponding expected frequency when it is computed.

Table 5.1
$A \times B \times C$ Combination Table

Description	$A \times B \times C$ Combination	fo	fe
Black border, red chin, blue bars	$A_1 B_1 C_1$	12	9
Black border, red chin, no blue bars	$A_1 B_1 C_2$	8	9.24
Black border, no red chin, blue bars	$A_1 B_2 C_1$	14	9.75
Black border, no red chin, no blue bars	$A_1 B_2 C_2$	4	10
No black border, red chin, blue bars	$A_2 B_1 C_1$	5	8.76
No black border, red chin, no blue bars	$A_2 B_1 C_2$	11	9
No black border, no red chin, blue bars	$A_2 B_2 C_1$	6	9.5
No black border, no red chin, no blue bars	$A_2 B_2 C_2$	15	9.75

$$N = 75$$

2. Combine the information from the $A \times B \times C$ combination table to form three 2×2 tables, giving all two-way combinations of

independent variables A, B, and C (see Table 5.2). Each table includes the total number of measurements but combines them into $A \times B$, $A \times C$, and $B \times C$ arrays, respectively. In each case, this is done by assuming that the third factor does not exist and by combining the pairs of combinations that are identical except for the third factor. For example, to obtain the $A \times B$ array, begin with cell A_1B_1. This value is obtained from the $A \times B \times C$ combination table by summing the values for $A_1B_1C_1$ and $A_1B_1C_2$ (that is, by ignoring the differences in the third factor). When you have completed the four cells for each table, check to be sure that the total number of measurements equals N, the sum of the *fo*s.

Table 5.2
Two-Way Combinations of Input Dimensions A, B, and C

$A \times B$

Black Border: A

		Yes A_1	No A_2	
Red Chin: B	Yes: B_1	20	16	36
	No: B_2	18	21	39
		38	37	75

$A \times C$

Black Border: A

		Yes A_1	No A_2	
Blue Bars: C	Yes: C_1	26	11	37
	No: C_2	12	26	38
		38	37	75

$B \times C$

Red Chin: B

		Yes B_1	No B_2	
Blue Bars: C	Yes: C_1	17	20	37
	No: C_2	19	19	38
		36	39	75

D. COMPUTATIONS

1. Compute appropriate *fe*s for χ^2_{Total}.

$$fe_{A_1 B_1 C_1} = \frac{(fo_{A_1})(fo_{B_1})(fo_{C_1})}{(N)(N)}, \qquad fe_{A_1 B_1 C_2} = \frac{(fo_{A_1})(fo_{B_1})(fo_{C_2})}{(N)(N)},$$

and so on.

For each combination in the $A \times B \times C$ combination table, the expected frequency is computed by multiplying the appropriate marginals for A, B, and C and then dividing by N times N.[9] You must perform this operation for each combination, using the appropriate A, B, and C values in each computation. For example, for $fe_{A_1B_1C_1}$, go to the marginals of the $A \times B$, $A \times C$, and $B \times C$ arrays. A_1 is the marginal (the total value for A_1 across the other conditions) from either the $A \times B$ or the $A \times C$ table. B_1 is the marginal from the appropriate table, and C_1 is found in the same way. (There will be eight such computations in a $2 \times 2 \times 2\chi^2$.) Enter the *fe*s in the $A \times B \times C$ combination table next to the *fo*s.

Let A = border variable   so that A_1 = black border A_2 = no black border

B = chin variable B_1 = red chin B_2 = no red chin

C = bar variable C_1 = blue bars C_2 = no blue bars

$$fe_{A_1B_1C_1} = \frac{(38)(36)(37)}{(75)(75)} = \frac{50616}{5625} = 9 \qquad fe_{A_2B_1C_1} = \frac{(37)(36)(37)}{5625} = 8.76$$

$$fe_{A_1B_1C_2} = \frac{(38)(36)(38)}{5625} = 9.24 \qquad fe_{A_2B_1C_2} = \frac{(37)(36)(38)}{5625} = 9$$

$$fe_{A_1B_2C_1} = \frac{(38)(39)(37)}{5625} = 9.75 \qquad fe_{A_2B_2C_1} = \frac{(37)(39)(37)}{5625} = 9.5$$

$$fe_{A_1B_2C_2} = \frac{(38)(39)(38)}{5625} = 10 \qquad fe_{A_2B_2C_2} = \frac{(37)(39)(38)}{5625} = 9.75$$

CHECK: $\Sigma fe = \Sigma fo = N$

The sum of the expected frequencies must equal the sum of the observed frequencies, which must equal N, the total number of subjects.

$$9 + 9.24 + 9.75 + 10 + 8.76 + 9 + 9.5 + 9.75 = 12 + 8 + 14 + 4$$
$$+ 5 + 11 + 6 + 15 = 75$$

---

[9]Conceptually, the procedure involves multiplying the probability of being in a given row by the probability of being in a given column by the probability of being in a given slice, and multiplying that product by N. Thus for cell $A_1B_1C_1$, the *fe* is $(fo_{A_1}/N)(fo_{B_1}/N)(fo_{C_1}/N)$ (N). This equation simplifies to $[(fo_{A_1})(fo_{B_1})(fo_{C_1})]/[(N)(N)]$.

CHECK: All *fe*s are greater than five. OK!

2. Solve for the three two-way χ^2s. In each array, the letters a, b, c, and d are assigned to cells starting in the upper left-hand cell. The marginal frequencies are designated by $a + b$, $c + d$, $a + c$, and $b + d$.

$$\chi^2_{A \times B} = \frac{N(bc - ad)^2}{(a + b)(c + d)(a + c)(b + d)}$$

$$\chi^2_{A \times C} = \frac{N(bc - ad)^2}{(a + b)(c + d)(a + c)(b + d)}$$

$$\chi^2_{B \times C} = \frac{N(bc - ad)^2}{(a + b)(c + d)(a + c)(b + d)}$$

For each computation, be sure to select the data from the corresponding 2 $\times$ 2 matrix. For each χ^2: multiply the observed frequency of a by that of d. Subtract this from the product of the observed frequencies of cells b and c. Square the outcome and multiply this value by N. Divide this product by the product of the four marginal frequencies. (These computations are described in detail in Section D, *Computations*, of the analysis for the two-way χ^2 earlier in this chapter.)

NOTE: There is no correction for continuity used in the multilevel χ^2.

CHECK: $a + b + c + d = N$

$$\chi^2_{A \times B} = \frac{75(288 - 420)^2}{(38)(37)(36)(39)} = .66$$

CHECK: $20 + 16 + 18 + 21 = 75 = N$

$$\chi^2_{A \times C} = \frac{75(132 - 676)^2}{(38)(37)(37)(38)} = 11.23$$

CHECK: $26 + 11 + 12 + 26 = 75 = N$

$$\chi^2_{B \times C} = \frac{75(380 - 323)^2}{(36)(37)(38)(39)} = .123$$

CHECK: $17 + 20 + 19 + 19 = 75 = N$

3. Solve for χ^2_{Total}.

$$\chi^2_{Total} = \frac{(fo_{A_1B_1C_1} - fe_{A_1B_1C_1})^2}{fe_{A_1B_1C_1}} + \frac{(fo_{A_1B_1C_2} - fe_{A_1B_1C_2})^2}{fe_{A_1B_1C_2}}$$

$$+\ldots + \frac{(fo_{A_2B_2C_2} - fe_{A_2B_2C_2})^2}{fe_{A_2B_2C_2}}$$

For each three-way combination: take the observed frequency, subtract the expected frequency (#1), square the outcome, and divide it by the expected frequency. Then sum the terms for all eight possible combinations.

$$\chi^2_{Total} = \frac{(12-9)^2}{9} + \frac{(8-9.24)^2}{9.24} + \frac{(14-9.75)^2}{9.75}$$

$$+ \frac{(4-10)^2}{10} + \frac{(5-8.76)^2}{8.76} + \frac{(11-9)^2}{9}$$

$$+ \frac{(6-9.50)^2}{9.50} + \frac{(15-9.75)^2}{9.75}$$

$$= \frac{9}{9} + \frac{1.54}{9.24} + \frac{18.06}{9.75} + \frac{36}{10} + \frac{14.14}{8.76} + \frac{4}{9} + \frac{12.25}{9.50} + \frac{27.56}{9.75}$$

$$= 1 + .17 + 1.85 + 3.6 + 1.61 + .44 + 1.29 + 2.83 = 12.79$$

CHECK: $\Sigma(fo - fe) = 0$

The sum of the observed frequencies minus the expected frequencies must equal 0.

$$(12-9) + (8-9.24) + (14-9.75) + (4-10) + (5-8.76)$$
$$+ (11-9) + (6-9.50) + (15-9.75) = 0$$

4. $\chi^2_{A \times B \times C} = \chi^2_{Total} - \chi^2_{A \times B} - \chi^2_{A \times C} - \chi^2_{B \times C}$
Solve for $\chi^2_{A \times B \times C}$ by subtracting from χ^2_{Total} (#3) the values for each of the three two-way χ^2s (#2).

$$12.79 - .66 - 11.23 - 123 = .78$$

5. Evaluate separately the significance of each of the three two-way χ^2s and that of the one three-way χ^2. The tabled value for χ^2 with 1 *df* and $\alpha = .05$ is 3.84. It is not necessary to evaluate χ^2_{Total}.

6. For the three two-way χ^2s:
   a. If $\chi^2_{Computed} \geq \chi^2_{Tabled}$, then there is a significant association between those two independent variables. Reject the null hypothesis.
   b. If $\chi^2_{Computed} < \chi^2_{Tabled}$, then the association between those two independent variables is not significant. Do not reject the null hypothesis.

For $\chi^2_{A \times B}$, $.66 < 3.84$. Therefore there is not a significant association between black borders and red chins. Do not reject the null hypothesis.

For $\chi^2_{A \times C}$, $11.23 > 3.84$. Therefore there is a significant association between black borders and blue bars. Reject the null hypothesis.

For $\chi^2_{B \times C}$, $.123 < 3.84$. Therefore there is not a significant association between red chins and blue bars. Do not reject the null hypothesis.

---

7. For the one three-way χ^2:
   a. If $\chi^2_{Computed} \geq \chi^2_{Tabled}$, then there is a significant association among the three independent variables. (Reject the null hypothesis.) Such an association will usually take one of two forms:
      (1) There is a strong association between two of the independent variables for one level of the third independent variable, but there is no association between the first two independent variables at the other level of the third.
         or
      (2) There is an association between two of the independent variables at one level of the third and an opposite association between the first two at the other level of the third.
   b. If $\chi^2_{Computed} < \chi^2_{Tabled}$, then the association among the three independent variables is not significant.

For $\chi^2_{A \times B \times C}$, $.78 < 3.84$. Therefore there is not a significant association among black borders, red chins, and blue bars.

E. INTERPRETATION OF THE EXAMPLE

1. *Statistical Conclusion*

   a. *For the three two-way χ^2.* If these observations were drawn randomly from a population in which there was no association between border color and chin color or between chin color and bar color, cell frequencies as discrepant as these would occur more often than 5% of the time by chance. Therefore these two null hypotheses should *not* be rejected. However, if these observations were drawn randomly from a population in which there was no association between border color and bar color, cell frequencies as discrepant as these would occur less than 5% of the time by chance. Therefore the null hypothesis should be rejected.

   b. *For the one three-way χ^2.* If these observations were drawn randomly from a population in which there was no association among border color, chin color, and bar color, cell frequencies as discrepant as these would occur more often than 5% of the time by chance. Therefore the null hypothesis should *not* be rejected.

2. *Researcher's Conclusion*

   The only two characteristics that are associated with one another are the presence or absence of the black border and the presence or absence of blue bars [from $\chi^2_{A \times C}$]. Fish that have blue bars also tend to have black borders. However, the color of a guppy's chin tells you nothing about whether that guppy will have either a black border or a blue bar [from $\chi^2_{A \times B}$ and $\chi^2_{B \times C}$]. Moreover, the presence of a third characteristic does not affect the degree of association between the other two [from $\chi^2_{A \times B \times C}$]. For example, the association between blue bars and black borders is unaffected by whether or not a fish has a red chin.

F. FURTHER PROCEDURES

1. Interpretation: see Chapter 9, *Introduction* and Section A.
2. Specific comparisons: in some cases, the researcher may wish to test the marginal frequencies from one or more of the two-way tables or the three-way table against some theoretical expectation (see p. 59). No other specific comparisons are necessary.
3. Strength-of-association measure: see Chapter 11, *Introduction* and Section A.

Description:

Test of association
Three independent
    variables
More than two levels
    of any variable
Between-subjects
Specific compari-
    sons available
Strength of asso-
    ciation available

THREE-WAY CHI SQUARE: ($a \times b \times c \; \chi^2$)

THREE INDEPENDENT VARIABLES

WITH MORE THAN TWO LEVELS OF

ANY VARIABLE

Follow through the steps below.

- *Have you read the* Limitations and Exceptions *section?*

- *Are all of these assumptions met for your study? If not, do not continue.*

- *This analysis is not appropriate if any* fes *are less than two. See p. 62.*

- *The three-way χ^2 with more than two levels of any independent variable is merely an expansion of the three-way ($2 \times 2 \times 2$) χ^2 detailed on p. 79. This design is not covered in complete detail because of the numerous computational steps involved.*

A. RECORDING THE DATA

1. Construct a table listing all possible $A \times B \times C$ combinations (in *Description* column). (For an example, see *Recording the Data*, p. 81.) For each combination, record the corresponding observed frequency (*fo* column). Record the expected frequencies (*fe* column) when they are computed.

2. Combine the information from the $A \times B \times C$ combination table to form three tables giving all two-way combinations of the independent variables A, B, and C. Each table includes the total number of measurements but combines them into $A \times B$, $A \times C$, and $B \times C$ arrays, respectively. (For an example, see Table 5.2.)

B. THE MAJOR COMPUTATIONAL STEPS

1. Compute $\chi^2_{A \times B}$
2. Compute $\chi^2_{A \times C}$ for $df = 1$, see p. 79
3. Compute $\chi^2_{B \times C}$ for $df > 1$, see p. 74 for computations

4. Compute χ^2_{Total}. See p. 80 for computations.

5. Obtain $\chi^2_{A \times B \times C}$ by subtraction. $\chi^2_{A \times B \times C} = \chi^2_{Total} - \chi^2_{A \times B} - \chi^2_{A \times C} - \chi^2_{B \times C}$.

If $\chi^2_{A \times B}$, $\chi^2_{A \times C}$, or $\chi^2_{B \times C}$ involve only 1 df, the computing formulas presented for the $2 \times 2 \times 2$ χ^2 (p. 79) should be followed. If they involve more than 1 df, the formula presented for the $a \times b$ χ^2 (p. 74) should be used to compute the two-way χ^2. Since all computations (except for the two-way χ^2 with df greater than 1 and the interpretations are identical to those presented for the $2 \times 2 \times 2$ χ^2, you may turn to p. 79 and follow the steps presented there.

C. FURTHER PROCEDURES

1. Interpretation: see Chapter 9, *Introduction* and Section A.
2. Specific comparisons: see Chapter 10, *Introduction* and Section A.
3. Strength-of-association measure: see Chapter 11, *Introduction* and Section A.

Description:

Test of association
Two independent
  variables
Two or more levels
  of either
  independent
  variable
"Repeated
  measures"
Specific compari-
  sons available
Strength of asso-
  ciation available
McNemar's χ^2_c may
  be performed
  on these data

"REPEATED-MEASURES" CHI SQUARE

TEST OF ASSOCIATION

Follow through the steps below.

● *Have you read the* Limitations and Exceptions *section?*

● *Are all of these assumptions met for your study? If not, do not continue.*

● χ^2 *may be used on "repeated measures" only when it is possible to set up the analysis so that subjects are counted only once (see example). When this can be done, the χ^2 provides a test of the null hypothesis that there is no association between the two classifications.*

Computations

For repeated measures having two levels of each independent variable, follow *Computations* and *Further Procedures* for the $2 \times 2 \chi^2$.

For repeated measures having more than two levels of either independent variable, follow *Computations* and *Further Procedures* for the two-way $a \times b \chi^2$.

A. THE EXAMPLE

1. A marine biologist trained largemouth bass to negotiate an underwater maze for a minnow reward. After the bass had 30 trials on the maze, the biologist noticed that 25 of his bass had solved the maze but that 14 of them were still making a large number of errors. He wondered whether the same bass that had solved his maze would be more or less likely to solve a different type of underwater maze, which he had used in previous research. He then gave all 39 bass 30 trials on the older maze to determine whether there was an association between their ability to solve Maze 1 and their ability to solve Maze 2 (the old maze).
2. *Null Hypothesis*
   a. *General.* In the population being sampled, an individual's level on Factor 1 is independent of his level on Factor 2.
   b. *Specific.* In the population being sampled, the probability that a largemouth bass is able to solve Maze 2 is independent of his ability to solve Maze 1, and vice versa.
3. *Steps through the Branching Program for This Example*
   a. *Statistical technique needed:* **test of significance**
   b. *Kind of data:* **frequency.** The number of bass that solved the maze and the number of bass that did not solve the maze are counted.

c. *Between, within, mixed:* "**repeated measure.**" Each bass has solved or not solved Maze 1, and each bass has solved or not solved Maze 2. However, in a 2 × 2 contingency table, a tally for each bass will occur in only one of the cells. Thus the scores are independent and the analysis is legitimate.

d. *Number of independent variables:* **two.** Performance on Maze 1 and performance on Maze 2

e. *Number of levels of each independent variable:* **performance on Maze 1 has two levels:** solved or not solved. **Performance on Maze 2 has two levels:** solved or not solved.

B. RECORDING THE DATA

Frequencies are recorded in an *a* × *b* table (see below).

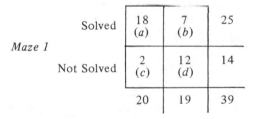

C. COMPUTATIONS

Since each independent variable has two levels, the computing formula for 2 × 2 χ^2 is used. (For more than two levels of either variable, the computations for the *a* × *b* χ^2, p. 74, are used.)

1. $$\chi^2 = \frac{N(|bc - ad| - N/2)^2}{(a + b)(c + d)(a + c)(b + d)} = \frac{39(|14 - 216| - 39/2)^2}{(20)(19)(25)(14)}$$

$$= 9.76$$

CHECK: $a + b + c + d = 18 + 7 + 2 + 12 = 39 = N$

2. $df = 1$. Tabled value of χ^2 with 1 df and $\alpha = .05$ is 3.84.

3. $9.76 > 3.84$. Therefore there is a significant association between solutions on Maze 1 and solutions on Maze 2. Reject the null hypothesis.

D. INTERPRETATION OF THE EXAMPLE

1. *Statistical conclusion.* If these observations were drawn randomly from a population in which there were no associations between learning to solve Maze 1 and learning to solve Maze 2, cell frequencies as discrepant as these would occur less than 5% of the time by chance. Therefore the null hypothesis should be rejected.

2. *Researcher's conclusion.* The significant association between solutions to Maze 1 and solutions to Maze 2 indicates that bass who solve Maze 1 are more likely to solve Maze 2 and that bass who fail to solve Maze 1 are less likely to solve Maze 2.

McNEMAR'S χ_c^2 (2 × 2 χ^2)   "REPEATED-MEASURES"   CHANGE TEST

Description:

Change test
Two independent variables
Two levels of each variable
"Repeated measures"
Specific comparisons: not necessary
Strength of association: not appropriate
Test of association (p. 89) may be performed on these data

Follow through the steps below.

● *Have you read the* Limitations and Exceptions *section?*

● *Are all of these assumptions met for your study? If not, do not continue.*

Frequency Data 93

• *If you are experienced with computations for χ_c^2, go directly to the computing formula below and proceed. If you require step-by-step guidance, begin with Section B, The Example. Remember, χ_c^2 does not test the null hypothesis that there is no association between classification at Time 1 and Time 2. That null hypothesis is tested by the repeated-measures χ^2, p. 89. χ_c^2 tests only the null hypothesis that the number of changers in one direction does not differ from the number of changers in the other direction. (See Footnote 2, p. 60.)*

A. COMPUTING FORMULA

$$\chi_c^2 = \frac{(|b-c|-1)^2}{b+c}$$

where b = the number of changers in one direction and
c = the number of changers in the other direction.

B. THE EXAMPLE

1. Fifty-one people were asked whether they believed there was convincing evidence for psychic phenomena. Twenty-three responded "believe" and 28 responded "don't believe." One year later, the same 51 people were asked the same question. This time, 36 responded "believe" and 15 responded "don't believe." Cross-classifying the people on their responses at Time 1 as compared to Time 2, we find that 20 of the people who initially responded "believe" still responded "believe," and 3 people who initially responded "believe" now responded "don't believe." Of those who initially responded "don't believe," 12 still responded "don't believe" and 16 now responded "believe."
2. *Null Hypothesis*
   a. *General.* In the population being sampled, the expected proportion in Category 1 on the first occasion is the same as the proportion in Category 1 on the second occasion. $H_0 : p_1 = p_2$.
   b. *Specific:* In the population being sampled, the same proportion of people today as a year ago would respond that they believe that there is convincing evidence for psychic phenomena.
3. *Steps through the Branching Program for This Example*
   a. *Statistical technique needed:* **test of significance**
   b. *Kind of data:* **frequency.** The number of persons who classified themselves as believing or not believing in psychic phenomena is counted.

c. *Between, within, mixed:* "**repeated measure.**" Each person is categorized as believing or not believing at Time 1 and as believing or not believing at Time 2. However, in the 2 X 2 contingency table, a tally for each person will occur in only one of the cells. Thus the scores are independent and the analysis is legitimate.

d. *Number of independent variables:* **two.** Classification at Time 1 and classification at Time 2.

e. *Number of levels of each independent variable:* **classification at Time 1 has two levels:** believing and not believing. **Classification at Time 2 has two levels:** believing and not believing.

C. RECORDING THE DATA

Record the data in a 2 X 2 matrix (see below), counting those who remain in the same classification from Time 1 to Time 2 in the upper-left and lower-right cell and those who change classifications in the upper right and lower-left cells.

Time 2

		Believe	Don't Believe
Time 1	Believe	20(a)	3(b)
	Don't Believe	16(c)	12(d)

D. COMPUTATIONS

1.
$$\chi_c^2 = \frac{(|b - c| - 1)^2}{b + c}$$

Solve for χ_c^2 by subtracting the number of changers in one direction from the number of changers in the other direction. Drop the sign, reduce this value by one, and square. Divide by the sum of the number of changers in one direction and the number of changers in the other direction.

$$\chi_c^2 = \frac{(|3 - 16| - 1)^2}{3 + 16} = \frac{(13 - 1)^2}{19} = \frac{144}{19} = 7.58$$

2. Tabled value of χ^2 with 1 df and $\alpha = .05$ is 3.84.

3. a. If χ_c^2 Computed $\geq \chi^2$ Tabled, the number of changers in one direc-

tion is significantly different from the number of changers in the other direction. Reject the null hypothesis.

b. If $\chi^2_{c\ \text{Computed}} < \chi^2_{\text{Tabled}}$, the number of changers in one direction does not differ significantly from the number of changers in the other direction. Do not reject the null hypothesis.

7.58 > 3.84. Therefore the number of changers in one direction is significantly different from the number of changers in the other direction.

E. INTERPRETATION OF THE EXAMPLE

1. *Statistical conclusion.* If these observations were drawn randomly from a population in which the proportions in cells *b* and *c* were equal, cell frequencies as discrepant as these would occur less than 5% of the time by chance. Therefore the null hypothesis should be rejected.

2. *Researcher's conclusion.* A significantly larger number of people changed their label from nonbelievers to believers than from believers to nonbelievers during the time period studied.

F. FURTHER PROCEDURES

1. Interpretation: see Chapter 9, *Introduction* and Section A.
2. Specific comparisons: not necessary.
3. Strength-of-association measure: not appropriate.

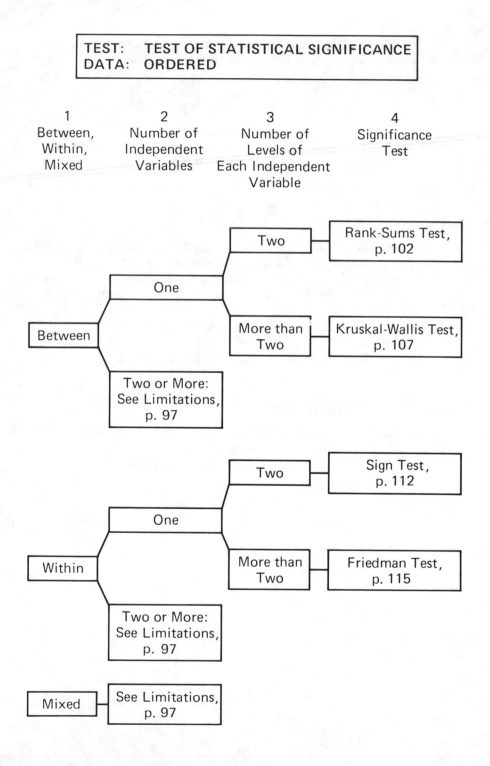

TEST: TEST OF STATISTICAL SIGNIFICANCE
DATA: ORDERED

| 1 Between, Within, Mixed | 2 Number of Independent Variables | 3 Number of Levels of Each Independent Variable | 4 Significance Test |

Between
- One
  - Two → Rank-Sums Test, p. 102
  - More than Two → Kruskal-Wallis Test, p. 107
- Two or More: See Limitations, p. 97

Within
- One
  - Two → Sign Test, p. 112
  - More than Two → Friedman Test, p. 115
- Two or More: See Limitations, p. 97

Mixed — See Limitations, p. 97

CHAPTER SIX

ORDERED DATA

LIMITATIONS AND EXCEPTIONS

No simple significance tests for ordered data with more than one independent variable are commonly used by psychologists. Therefore all such tests have been omitted from this book. (Because of this limitation, we have included no ordered-data tests for mixed designs, since by definition there are at least two independent variables in a mixed design.) If you have ordered data and more than one independent variable, consider one of the following:

1. For ranks derived from numerical scores, you may assume that the assumptions underlying tests designed for score data can be met, and you may treat the data as score data. This technique is widely used, although statisticians disagree on its legitimacy.
2. The ordered data may be converted to frequency data by recording the frequency of the people who fall into various categories. The frequency data may then be analyzed by the use of a multi-classification χ^2.

There are two basic conditions under which a researcher may have ordered data and may therefore wish to use one of the tests of significance included in this chapter. First: no numerical scores were obtained for any subjects, but the researcher was able to order the subjects along some dimension of interest. In this case, the *raw data* would be ordered data (ranks) and the computing formulas may be used directly for most analyses. Second (and more common): the researcher obtained numerical scores for his subjects but felt that they were not precise or that the assumptions that underlie tests for score data were violated. In this latter case, the researcher *orders* the subjects on the basis of their scores and performs an analysis on the ordered data.

In Chapter 5 we present one basic statistical test, χ^2, and show that most designs for frequency data can be encompassed by extensions of this single statistic. Similarly, in Chapters 7 and 8, analysis of variance is expanded to cover a wide variety of designs for score data. Unfortunately, there is no single general technique for analyzing ordered data. Statistical tests appropriate for ordered data cover fewer research designs than do tests for frequency or score data. Both the statisticians' lack of interest in developing such tests and the mathematical complexity of developing them contribute to the relatively primitive state of analyses for ordered data.

There are no simple tests commonly used for ordered data when there are more than one independent variable in the study. Therefore, in this chapter we have included only those significance tests appropriate for research designs with a single independent variable. These tests are appropriate for four types of designs: (1) between-subjects design, two levels of the independent variable; (2) between-subjects design, more than two levels of the independent variable; (3) within-subjects design, two levels of the independent variable; and (4) within-subjects design, more than two levels of the independent variable. We have included no complex designs for ordered data (since, by definition, they require more than one independent variable).

Each of the tests in this chapter works in a relatively simple way. If there is no relationship between the independent and the dependent variables, the various groups in the experiment should have essentially the same average rank. (In effect, the null hypothesis would be that the groups were drawn from populations whose distributions are exactly alike.) If the null hypothesis is false and there is an association between the independent and the dependent variables, the sums of ranks in the various groups should be quite discrepant from one another.

PARAMETRIC AND NONPARAMETRIC STATISTICS

Tests for ordered data may be used when scores do not meet the assumptions underlying and required for tests of score data (parametric assumptions). If such data are converted from score data to ordered data, only nonparametric assumptions need be met. Since statisticians still disagree on the relative merits of parametric and nonparametric techniques, we will discuss this problem briefly. Parametric techniques, such as the *t* test and the analysis of variance, estimate population parameters. They also make a number of assumptions about the populations from which the scores were drawn. Most importantly, they assume that (1) the samples were drawn at random from the populations under consideration, (2) the variances in the populations are homogeneous, and (3) the scores are normally distributed in the populations. Nonparametric techniques, such as the ones presented in this chapter, make much weaker assumptions. The usual nonparametric assumptions are that (1) the samples were drawn at random from the populations under consideration and (2) the scores underlying the ranks form a continuous distribution.

At one extreme, some statisticians argue that the assumptions underlying parametric techniques are virtually never justified for behavioral science research and, as a consequence, that only nonparametric techniques should be used. At the other extreme, many statisticians argue that violations of these assumptions do not seriously impair the usefulness of parametric techniques and that they should always be used because of their greater power. Empirical studies do indicate that violations of the assumptions of equal variances and normal distribution of scores do not severely affect the outcome of the analysis of variance or of the *t* test. Analysis of variance can be used effectively in many situations in which these assumptions are violated, if certain precautions are taken. This topic is discussed more fully in Chapter 7.

We are often more concerned with the problem of imprecise measurement techniques than with the problem of assumptions. Analysis of variance assumes that equal numerical intervals between scores correspond to equal empirical intervals in the underlying behaviors being measured. There is a lively argument among statisticians involving the legitimacy of using analysis of variance on scores that fail to meet this equal-interval assumption.[1] When extremely imprecise measuring techniques are used, it is probably better to rank the data (converting from score data) and use one of the techniques presented in this chapter. A judgment is involved that each individual researcher must make. However, it must be emphasized that, when there is more

---

[1] For a full discussion of the problems involved in relating statistical tests to underlying measurement dimensions, see Kirk (1972), pp. 47–80.

than one independent variable, none of the present techniques is appropriate. The researcher must either use analysis of variance or convert the data to frequencies and apply a χ^2.

NOTE ON ASSIGNING RANKS AND THE PROBLEM OF TIES

If you have numerical scores that you wish to change to ranks, list them in ascending or descending order (specific analyses indicate whether all scores, scores within a condition, or scores for a specific subject are ranked). Assigned ranks go from 1 to N (the total number of scores ranked). Whether 1 is assigned to the highest or to the lowest numerical rank does not matter in any of the tests given here.

You will simplify problems of interpretation if you assign ranks in a way that is comfortable and natural to you. For example, if a score represents self-esteem, decide whether a low rank (1, 2, 3) or a high rank (15, 16, 17) better represents high self-esteem. Adopt whichever is more comfortable for you. An apt choice reduces the number of translations between the meaning of the assigned rank and the scores, and it decreases the probability of errors in interpretation.

Tied numerical scores and ranks.[2] If you have numerical scores that are equal, assign equal ranks to them. For example, if *two* scores that would be ranks 6 and 7 are tied, they are each given the rank of 6.5. It is important to remember that the *next* score is given the rank 8. If three scores that would be ranks 3, 4, and 5 are tied, they are all given the rank 4. It is important to remember that the *next* score is given the rank 6. Table 6.1 illustrates this procedure. Ties among ranks are handled in an exactly comparable fashion. If there is a two-place tie for the third position, the tied subjects would each receive the rank 3.5, and the next subject would receive the rank of 5.

CHECK: Count the total number of scores (N). Be sure that your highest rank is assigned this value (unless there is a tie for the highest rank). If this check is not verified, there is an error. A good place to begin checking is with your assignment of tied ranks.

---

[2]We have presented the simplest of several methods for dealing with tied ranks. For a discussion of alternative methods, see Wike (1971), pp. 107–109.

Table 6.1. Assignment of Ranks to Tied Scores

Scores	Ranks	
3	1	
8	2	
10	3	
11	4.5	Ranks 4 and 5 would be assigned to these two scores if they were untied. The average rank, 4.5, is assigned to each.
11	4.5	
12	6	Rank 6, not 5, is assigned to the value following rank 4.5.
13	8	Ranks 7, 8, and 9 would be assigned to untied values. The average of these ranks, rank 8, is assigned to each.
13	8	
13	8	
21	10	Rank 10, not rank 9, is assigned to the value following this tie.
63	11	CHECK: The rank 11 corresponds to N, the total number of scores.

NOTE ON THE ANALYSES AND WORKED EXAMPLES

Each analysis is presented with spelled-out step-by-step procedures. For each analysis there is a worked example. The example (1) begins with a description of the study, (2) presents an appropriate null hypothesis, (3) indicates the steps through the branching program, (4) goes step by step through the computations, and (5) presents the statistical and research outcomes of the study.

Since later computations refer back to steps by number (for example, "#4 less #5"), you will save time if you list the products from each step by number as you proceed. In addition, you can locate and correct errors in your computations more easily if you have recorded your intermediate steps systematically.

Accompanying each analysis is a brief description that lists in a summary form when the analysis may be used, special requirements, and so on.

Each analysis provides a reference to chapters on interpretation, specific comparison, and strength of association. Each example includes both a statistical conclusion and a journal-style interpretation (researcher's conclusion). We have not included either the specific-comparison tests or the strength-of-association measures for these examples.

RANK-SUMS TEST

Description:

Two levels of the independent
    variable
Between-subjects
Considerable power for conven-
    tional differences between
    groups
Strength-of-association measure
    available

Follow through the steps below.

● *Have you read the* Limitations and Exceptions *section?*

● *Are all of these assumptions met for your study? If not, do not continue.*

● *If you have fewer than eight scores in either group, the present procedures can't be used.*[3]

● *If you are experienced with computations for the rank-sums test, go directly to the computing formula below. If you require step-by-step guidance, begin with Section B,* The Example.

---

[3]If either n_1 or n_2 is less than eight, the Z test is inappropriate. A table of the exact probabilities associated with the rank-sums test for small samples is available in Senders (1958), pp. 541–542.

A. COMPUTING FORMULA

$$Z = \frac{2T_i - n_i(N + 1)}{\cdot\sqrt{\dfrac{n_1 n_2 (N + 1)}{3}}}$$

where T_i = the sum of the ranks for either group,
n_i = the number of Ss in the group used to get T_i,
N = the total number of Ss,
n_1 = the number of Ss in Group 1, and
n_2 = the number of Ss in Group 2.

The critical value of Z at the .05 level is ± 1.96. If your obtained value is greater than $+ 1.96$ or less than $- 1.96$, your difference is significant. (Note that this means that numbers larger than 1.96, regardless of sign, are significant.)

B. THE EXAMPLE

1. A group of 21 Labrador puppies were randomly assigned to receive either a special dietary supplement (DS group) or to receive normal feeding (NF group). Eleven puppies were in the DS group, 10 in the NF group. At the end of six months, a panel of judges who were unaware of the type of feeding that the dogs received ranked all of the dogs on the basis of their general appearance and vitality. The best dog of the group was ranked 1, the second best 2, and so on.

   NOTE: $N > 8$ for each group, so special procedures for small samples are not required.

2. *Null Hypothesis*
   a. *General.* These two samples are drawn from populations that are identically distributed.
   b. *Specific.* In the populations being sampled, the judged general appearance and vitality of dogs is distributed in the same way for dogs that receive a special dietary supplement as for those that receive normal feeding.
3. *Steps through the Branching Program for This Example*
   a. *Statistical technique needed:* **test of significance**
   b. *Kind of data:* **ordered.** Dogs are ranked on appearance. (If scores were obtained, they should be converted to ranks.)
   c. *Between, within, mixed:* **between.** Each dog received only one kind of feed and is ranked only once.

d. *Number of independent variables:* **one**. Feeding type
e. *Number of levels of independent variable:* **two**. Normal or supplement

C. RECORDING THE DATA

Record the data in a table (see Table 6.2) having the following headings:

1. *Scores*. If you have numerical scores, list them in ascending (or descending) order.
2. *Ranks*. Assign a rank to each score. If you have no scores, list the ranks in ascending order. Assign tied ranks to tied scores. (See *Assigning Ranks and the Problem of Ties*, p. 100).
3. *Groups*. Identify the group to which each rank belongs.
4. *Group 1*. Write down all of the ranks for subjects in Group 1.
5. *Group 2*. Write down all of the ranks for subjects in Group 2.

Table 6.2. Data Table: Rank-Sums Test

Rank	Group	DS Group	NF Group
1	DS	1	
2	DS	2	
3.5	DS	3.5	
3.5	DS	3.5	
5	NF		5
6	DS	6	
7	DS	7	
8	DS	8	
9	NF		9
10	DS	10	
11.5	DS	11.5	
11.5	DS	11.5	
13	NF		13
14	NF		14
15	NF		15
16	NF		16
17	NF		17
18	DS	18	
19	NF		19
20	NF		20
21	NF		21
		$T_1 = 82$	$T_2 = 149$
		$n_1 = 11$	$n_2 = 10$

NOTE: Since there are no numerical scores in the example, there are only four headings in the data table.

D. COMPUTATIONS

1. Obtain n_1 and n_2: count the number of subjects in Group 1. Then count the number of subjects in Group 2.

$$n_1 = 11, \qquad n_2 = 10$$

CHECK: $n_1 + n_2 = N$

$$11 + 10 = 21 \qquad \text{OK!}$$

2. Obtain T_1 and T_2: sum the ranks for Group 1. Then sum the ranks for Group 2.

$$T_1 = 82, \qquad T_2 = 149$$

CHECK: $T_1 + T_2$ $= 82 + 149 = 231$ $= \dfrac{N(N + 1)}{2}$

$$= \frac{21(22)}{2} = \frac{462}{2} = 231. \qquad \text{OK!}$$

3. a. Select a T_i and n_i: select either T_1 or T_2 to be T_i. If T_1 is selected, use n_1 as n_i. If T_2 is selected, use n_2 as n_i. (The choice of total affects the sign of the statistic Z, but not the size.)

We have selected T_1 to be T_i; therefore T_i will be 82 and n_i will be 11.

b. Solve for: $2T_i - n_i(N + 1)$.

$$2(82) - 11(22) = 164 - 242 = -78$$

4. Solve for: $\sqrt{\dfrac{n_1 n_2 (N + 1)}{3}}$

$$\sqrt{\frac{10(11)(22)}{3}} = \sqrt{\frac{2420}{3}} = \sqrt{806.67} = 28.4$$

5. Find $Z = \dfrac{2T_i - n_i(N + 1)}{\sqrt{\dfrac{n_1 n_2 (N + 1)}{3}}}$. Divide # 3b by # 4 to obtain Z.

$$\frac{-78}{28.4} = -2.75$$

6. The critical value of Z at the .05 level is ± 1.96.
7. a. If the obtained Z is greater than $+ 1.96$ or less than $- 1.96$, the difference between the two groups is significant. (Note that this means that numbers larger than 1.96, regardless of sign, are significant.) Reject the null hypothesis.
   b. If the obtained Z is less than $+ 1.96$ but greater than $- 1.96$, the difference between the two groups is not significant. (Note that this means that numbers smaller than 1.96, regardless of sign, are not significant.) Do not reject the null hypothesis.

$- 2.75$ is less than $- 1.96$ (the absolute value is larger); therefore the two groups are significantly different. Reject the null hypothesis.

E. INTERPRETATION OF THE EXAMPLE

1. *Statistical conclusion*. If these observations were drawn randomly from identically distributed populations, sums of ranks as discrepant as these would occur less than 5% of the time by chance. Therefore the null hypothesis should be rejected.
2. *Researcher's conclusion*. (Remember that low ranks indicate vital dogs with good appearance.) Dogs receiving the dietary supplement were judged significantly better in appearance and vitality than dogs receiving a normal diet.

F. FURTHER PROCEDURES

1. Interpretation: see Chapter 9, *Introduction* and Section B.
2. Specific comparisons: no test is necessary.

3. Strength-of-association measure: see Chapter 11, *Introduction* and Section B.

	Description:
	a levels of variable
	For $a = 2$, see rank-sums test
KRUSKAL-WALLIS TEST	Between-subjects
	Tests for differences among groups
	Both strength of association and specific comparisons possible

Follow through the steps below.

● *Have you read the* Limitations and Exceptions *section?*

● *Are all of these assumptions met for your study? If not, do not continue.*

● *If a = 3, and the number in any group is five or fewer, an exact test is necessary.*[4]

● *If you are experienced with computations for the Kruskal-Wallis test, go directly to the computing formula below. If you require step-by-step guidance, begin with Section B,* The Example.

A. COMPUTING FORMULA

$$H = \frac{12}{N(N + 1)} \sum_{A=1}^{a} \frac{T_A^2}{n_A} - 3(N + 1)$$

[4]Wike (1971) presents tables for the exact test for Kruskal-Wallis, pp. 222–223.

where T_A = the sum of the ranks for each level of the independent
variable,

   n_A = the corresponding number of subjects for each T_A,

   N = the total number of subjects, and

   a = the number of levels of the independent variable.

H is evaluated as χ^2 with $a - 1$ *df* (see Appendix 1).

B. THE EXAMPLE

1. From the college population available to him, a sociologist
randomly selected seven married women, eight divorcees, and
nine single women. He was interested in determining whether
marital status was related to overall adjustment. He gave each
woman the Johnson Self-Administered Adjustment Test (JSAAT),
a test that he had constructed and was testing. Since he was not
at all sure how precise a measuring instrument the JSAAT was,
he decided to rank all of his subjects according to their scores
and perform an analysis on the ranks. He assigned a rank of 1
to the lowest score, and so on. Low scores indicated poor
adjustment.
2. *Null Hypothesis*
   a. *General.* These samples are drawn from populations that are
   identically distributed.
   b. *Specific.* In the population being sampled, the distribution of
   overall adjustment is the same for married women, divorcees,
   and single women.
3. *Steps through the Branching Program for This Example*
   a. *Statistical technique needed:* **test of significance**
   b. *Kind of data:* **ordered.** The original score data is converted to
   rank data.
   c. *Between, within, mixed:* **between.** Each woman is ranked
   only once.
   d. *Number of independent variables:* **one.** Marital status
   e. *Number of levels of the independent variable:* **three.** Married,
   divorced, single

C. RECORDING THE DATA

1. *If your raw data are ranks:* record the data directly under the
appropriate headings in a table with *a* columns (see Table 6.3).
2. *If you have numerical scores:*
   a. List the scores *for all subjects* in all conditions in ascending
   (or descending) order. Assign ranks from 1 to N to the scores.
   In case of tied scores, assign tied ranks. (See *Assigning Ranks*

and the Problem of Ties, p. 100, for a detailed treatment of tied scores.)

b. Enter the scores and the ranks for each group under the appropriate heading in the raw-data table.

Table 6.3. Data Table: Kruskal-Wallis Test

Married		Divorced		Single	
Score	Rank	Score	Rank	Score	Rank
18	3.5	12	1	18	3.5
28	8	16	2	21	5
32	9	37	11	26	6.5
46	13.5	40	12	26	6.5
52	16.5	46	13.5	33	10
62	20	52	16.5	51	15
63	21.5	61	19	53	18
		63	21.5	68	23
				70	24
$T_1 = 92.0$		$T_2 = 96.5$		$T_3 = 111.5$	
$n_1 = 7$		$n_2 = 8$		$n_3 = 9$	

D. COMPUTATIONS

1. Obtain the n_A s. Count the number of subjects in each group.

$$n_1 = 7, n_2 = 8, n_3 = 9$$

CHECK: $\sum_{A=1}^{a} n_A = N$

Sum the total number of subjects in each group. This must equal N.

$$7 + 8 + 9 = 24. \qquad \text{OK!}$$

2. Obtain the T_A s. Sum the ranks for each of the groups in the study.

$$T_1 = 92.0, T_2 = 96.5, T_3 = 111.5$$

CHECK: $\sum_{A=1}^{a} T_A$ $= 92.0 + 96.5 + 111.5 = 300$

$$= \frac{N(N+1)}{2} = \frac{24(25)}{2} = \frac{600}{2} = 300. \qquad \text{OK!}$$

3. Solve for: $\displaystyle\sum_{A=1}^{a} \frac{T_A^2}{n_A}$

Take the sums obtained in #2, square each value, and divide by the number of subjects in that group (the values from #1). Sum all of these values.

$$\frac{92^2}{7} + \frac{96.5^2}{8} + \frac{111.5^2}{9} = \frac{8464}{7} + \frac{9312.25}{8} + \frac{12432.25}{9}$$
$$= 1209.14 + 1164.03 + 1381.36 = 3754.53$$

4. Solve for: $\dfrac{12}{N(N+1)}$.

Divide 12 by the product of N times $N + 1$.

$$\frac{12}{24(25)} = \frac{12}{600} = .02$$

5. Solve for: $\dfrac{12}{N(N+1)} \displaystyle\sum_{A=1}^{a} \frac{T_A^2}{n_A}$.

Multiply #3 by #4.

$$.02(3754.53) = 75.09$$

6. Solve for: $3(N + 1)$.

$$3(25) = 75$$

7. $H = \left[\dfrac{12}{N(N+1)} \displaystyle\sum_{A=1}^{a} \frac{T_A^2}{n_A} \right] - [3(N+1)]$ Subtract: #5 − #6.

$$75.09 - 75 = .09$$

8. Obtain the critical value of H from the χ^2 table, Appendix 1, with $df = a - 1$.

> The critical value of χ^2 with $(3 - 1) = 2\,df$ and $\alpha = .05$ is 6.0.

9. a. If $H_{\text{Obtained}} \geq \chi^2_{\text{Tabled}}$, the differences among the groups are significant. Reject the null hypothesis.
   b. If $H_{\text{Obtained}} < \chi^2_{\text{Tabled}}$, the differences among the groups are not significant.[5] Do not reject the null hypothesis.

> .09 $<$ 6.0. Therefore the differences among the groups are not significant. Do not reject the null hypothesis.

E. INTERPRETATION OF THE EXAMPLE

1. *Statistical conclusion.* If these observations were drawn randomly from identically distributed populations, sums of ranks as discrepant as these would *not* occur less than 5% of the time by chance. Therefore the null hypothesis should *not* be rejected.
2. *Researcher's conclusion.* There is no basis for believing that marital status is related to overall adjustment, as measured by the JSAAT for this population.

F. FURTHER PROCEDURES

1. Interpretations: see Chapter 9, *Introduction* and Section B.
2. Specific-comparison tests: see Chapter 10, *Introduction* and Section B.
3. Strength-of-association measure: see Chapter 11, *Introduction* and Section B.

---

[5] If H_{Obtained} just misses significance and more than 10% of the ranks are tied, some statisticians recommend that a correction formula be applied. The correction formula is $C = 1 - [\sum\limits_{i=1}^{m} (t_i^3 - t_i)]/[N^3 - N]$ where $m =$ the number of sets of tied scores, $t_i =$ the number of scores in any set of tied scores, and $N =$ the total number of subjects. For example, suppose that you have three different ties: 6, 6, 6; 8, 8, 8; and 10, 10. All other scores in the array are unique. Then $m = 3$ (there are three different sets of ties), $t_1 = 3$, $t_2 = 4$, and $t_3 = 2$ (the number of tied scores in each of the three sets). (In this case, if $N = 24$, C would be $1 - .006 = .994$.) The test statistic becomes $H_C = H_{\text{Obtained}}/C$, which is tested as H.

Description:

Two levels of variable
Within-subjects
Easy to perform
Tests for differences between
   conditions
No strength-of-association
   measure available

THE SIGN TEST

Follow through the steps below.

● *Have you read the* Limitations and Exceptions *section?*

● *Are all of these assumptions met for your study? If not, do not continue.*

A. THE EXAMPLE

1. A nursery-school teacher observed a group of 28 children during a 30-minute play period. She evaluated the aggressiveness of each child on a scale from 1 to 10 in which 1 meant highly aggressive and 10 meant highly unaggressive. The children then watched a puppet show designed to elicit cooperation and suppress aggression. After the show, children were re-evaluated on the scale of aggressiveness.
2. *Null Hypothesis*
   a. *General.* The observations are drawn from a population in which a subject's score is as likely to increase as to decrease on two measurements.
   b. *Specific.* In the populations being sampled, children are as apt to be evaluated as more aggressive as to be evaluated as less aggressive after watching a puppet show designed to suppress aggression.
3. *Steps through the Branching Program for This Example*
   a. *Statistical technique needed:* **test of significance**
   b. *Kind of data:* **ordered**. The original score data has been transformed into signed data. (Signed data is a member of the general class of ordered data. A "+" is assigned if the measurement at Time 2 is greater than the measurement at Time 1. A "−" is assigned if the observation at Time 2 is smaller than the observation at Time 1. Assigning +s and −s is a weak form of ranking.)
   c. *Between, within, mixed:* **within**. Each child is measured twice.
   d. *Number of independent variables:* **one**. Time at which evaluations were made

e. *Number of levels of independent variable:* **two.** Before puppet show and after puppet show

B. RECORDING THE DATA

1. *If numerical scores were obtained:* list the scores for Measurement 1 and Measurement 2 for each subject. Assign "+" to changes in one direction. Assign "—" to changes in the other direction. *Eliminate from the sample the data of all subjects who did not change.*
2. *If no numerical scores were obtained:* assign "+" to changes in one direction and "—" to changes in the other direction. *Eliminate from the sample the data of all subjects who did not change.* See Table 6.4.

Table 6.4. Data Table: The Sign Test

Subject Number	Pretest	Post-Test	Change
1	2	4	+
2	4	5	+
3	1	3	+
4	3	2	—
5	5	4	—
6	6	8	+
7	1	5	+
8	4	5	+
9	3	7	+
10	2	5	+
11	2	6	+
12	8	8	0
13	9	6	—
14	5	6	+
15	3	4	+
16	6	9	+
17	7	10	+
18	2	6	+
19	4	5	+
20	9	6	—
21	5	7	+
22	4	6	+
23	3	2	—
24	3	6	+
25	6	4	—
26	4	5	+
27	2	3	+
28	1	3	+

"+ changes" indicate that the child became less aggressive;
"— changes" indicate that the child became more aggressive.

C. COMPUTATIONS

1. Count the total number of "+ changes."

21

2. Count the total number of "— changes."

6

3. The total number of *changers* (in both directions) is N. (Child
12 did not change and was eliminated, thus reducing N from
28 to 27.)

27

4. The *smaller* of the above totals will be the statistic R.

6

5. Obtain the critical value of R for the appropriate N at $\alpha = .05$
from Appendix 2. Remember that subjects who did not change
are not included in N.

Critical value of R at $\alpha = .05$ for N of 27 is 7.

6. a. If $R_{\text{Obtained}} \leq R_{\text{Tabled}}$, there is a significant change from
Measurement 1 to Measurement 2. Reject the null hypothesis.
   b. If $R_{\text{Obtained}} > R_{\text{Tabled}}$, there is not a significant change from
Measurement 1 to Measurement 2. Do not reject the null
hypothesis.

6 < 7. Therefore the number of changers is significant.
Reject the null hypothesis.

D. INTERPRETATION OF THE EXAMPLE

1. *Statistical conclusion.* If these observations were drawn randomly from a population in which judged aggressiveness was as likely to increase as to decrease, a proportion of decreases as great as this would occur less than 5% of the time by chance. Therefore the null hypothesis should be rejected.

2. *Researcher's conclusion.* [Remember that a "+ change" indicates that the child became less aggressive. There were more "+ changes" than "— changes"; therefore the change is in the direction of less aggressiveness.] Children are more likely to be evaluated as less aggressive than as more aggressive after watching the puppet show that discourages aggression.

E. FURTHER PROCEDURES

1. Interpretation: see Chapter 9, *Introduction* and Section B.
2. Specific-comparison tests: unnecessary.
3. Strength-of-association measure: there is no strength-of-association measure for the sign test.

Description:

a levels of variable
Within-subjects
Tests for differences among
    conditions
Both strength-of-association
    and specific-comparison
    tests possible

FRIEDMAN TEST

Follow through the steps below.

● *Have you read the* Limitations and Exceptions *section?*

● *Are all of these assumptions met for your study? If not, do not continue.*

● *If you have three levels of the independent variable (a = 3) and fewer than ten subjects per group, or if you have four levels of the independent variable (a = 4) and fewer than five subjects per group,* the present procedures can't be used.[6]

● *If you are experienced with computations for the Friedman test, go directly to the computing formula below. If you require step-by-step guidance, begin with Section B,* The Example.

A. COMPUTING FORMULA

$$\chi_r^2 = \frac{12}{(a)(s)(a + 1)} \sum_{A=1}^{a} T_A^2 - 3(s)(a + 1)$$

where s = the number of subjects,
         a = the number of measurements on each subject (the number of levels of the independent variable), and
         T_A = the sum of the ranks of each column.

χ_r^2 is evaluated as χ^2 with $(a - 1)$ df (see Appendix 1).

B. THE EXAMPLE

1. Ten subjects were randomly selected from a college population to serve on a mock jury. They were given a brief account of the case and were then asked to evaluate the defendant's guilt or innocence on a 7-point scale in which 1 indicated complete innocence and 7 indicated complete guilt. At two dramatic points in the mock trial, they were asked to re-evaluate the defendant on the same scale. Finally, after the summations were made, they evaluated the defendant for the fourth time.
2. *Null Hypothesis*
   a. *General*. In the population being sampled, the expected sum of ranks under each experimental condition is the same.
   b. *Specific*. In the population being sampled, the subjects' expected evaluation of the defendant's guilt or innocence is the same at all four times of evaluation.
3. *Steps through the Branching Program for This Example*
   a. *Statistical technique needed:* **test of significance**
   b. *Kind of data:* **ordered**. The original score data are converted to ranks.

[6]Wike (1971), p. 229, presents tables for the exact test for the Friedman test.

c. *Between, within, mixed:* **within.** Each subject is measured four times.

d. *Number of independent variables:* **one.** Sequence in the mock trial

e. *Number of levels of the independent variable:* **four.** Four measurements during the course of the mock trial

C. RECORDING THE DATA

1. Prepare a table with *a* headings appropriate to the treatment conditions (see Table 6.5).

Table 6.5. Data Table: Friedman Test Example

Subject Number	Measurement 1		Measurement 2		Measurement 3		Measurement 4	
	Score	Rank	Score	Rank	Score	Rank	Score	Rank
1	5	4	2	2	4	3	1	1
2	6	4	4	2	5	3	2	1
3	7	4	3	2.5	3	2.5	2	1
4	7	4	4	2	5	3	3	1
5	6	3.5	5	2	6	3.5	2	1
6	4	1	5	2.5	5	2.5	6	4
7	5	3	4	1.5	6	4	4	1.5
8	7	3.5	6	2	7	3.5	3	1
9	6	4	5	2.5	5	2.5	2	1
10	5	2.5	5	2.5	6	4	3	1
	$T_1 =$ 33.5		21.5		31.5		13.5	

2. Optional: enter the numerical scores, if any, under the appropriate headings for each subject.

3. a. The ranks for this analysis are the ranks of scores from 1 to *a* for an individual subject. That is, take Subject 1 and rank his scores from 1 to *a*, take Subject 2 and rank his scores from 1 to *a*, and so on.

   b. For each subject, enter the ranks under the heading for the appropriate condition.

D. COMPUTATIONS

1. Obtain the T_As. Sum the ranks for each of the measurements.

$$T_1 = 33.5, \quad T_2 = 21.5, \quad T_3 = 31.5, \quad T_4 = 13.5$$

CHECK: $\displaystyle\sum_{A=1}^{a} T_A$ $\boxed{33.5 + 21.5 + 31.5 + 13.5 = 100}$

$$= \frac{(a)(s)(a+1)}{2} \qquad \boxed{\begin{array}{l} = \dfrac{(4)(10)(5)}{2} = \dfrac{200}{2} \\ = 100. \qquad \text{OK!} \end{array}}$$

2. Solve for: $\displaystyle\sum_{A=1}^{a} T_A^2$. Square each T_A, then sum.

$$\boxed{\begin{array}{c} 33.5^2 + 21.5^2 + 31.5^2 + 13.5^2 = 1122.25 + 462.25 \\ + 992.25 + 182.25 = 2759 \end{array}}$$

3. Solve algebraically: $\dfrac{12}{(a)(s)(a+1)}$.

$$\boxed{\dfrac{12}{(4)(10)(5)} = \dfrac{12}{200} = .06}$$

4. Solve for: $\left[\dfrac{12}{(a)(s)(a+1)}\right]\left[\displaystyle\sum_{A=1}^{a} T_A^2\right]$. Multiply: #3 × #2.

$$\boxed{(.06)(2759) = 165.54}$$

5. Solve algebraically: $3(s)(a+1)$.

$$\boxed{3(10)(5) = 150}$$

6. Solve for: $\chi_r^2 = \left[\dfrac{12}{(a)(s)(a+1)}\right]\left[\displaystyle\sum_{A=1}^{a} T_A^2\right] - [3(s)(a+1)]$.

   Subtract: #4 − #5.

$$\boxed{165.54 - 150 = 15.54}$$

7. Obtain critical value of χ^2 from Appendix 1 with $(a - 1)\,df$ for $\alpha = .05$.

Critical value of χ^2 with $(4 - 1) = 3\,df$ for $\alpha = .05$ is 7.81.

---

8. a. If $\chi^2_{r\,\text{Obtained}} \geq \chi^2_{\text{Tabled}}$, the differences among the measurements are significant. Reject the null hypothesis.
   b. If $\chi^2_{r\,\text{Obtained}} < \chi^2_{\text{Tabled}}$, the differences among the measurements are not significant. Do not reject the null hypothesis.

$15.54 > 7.81$. Therefore the differences among the measurements are significant. Reject the null hypothesis.

E. INTERPRETATION OF THE EXAMPLE

1. *Statistical conclusion.* If these observations were drawn randomly from a population in which the subjects' expected evaluation of the defendant's guilt or innocence was the same at all four times of evaluation, then average ranks as discrepant as these would occur less than 5% of the time by chance. Therefore the null hypothesis should be rejected.
2. *Researcher's conclusion.* Subjects' evaluations of the defendant's guilt or innocence change over the course of the mock trial. [A specific-comparison test must be performed to determine which measurement differed from which other ones.]

F. FURTHER PROCEDURES

1. Interpretation: see Chapter 9, *Introduction* and Section B.
2. Specific-comparison tests: see Chapter 10, *Introduction* and Section B.
3. Strength-of-association measure: see Chapter 11, *Introduction* and Section B.

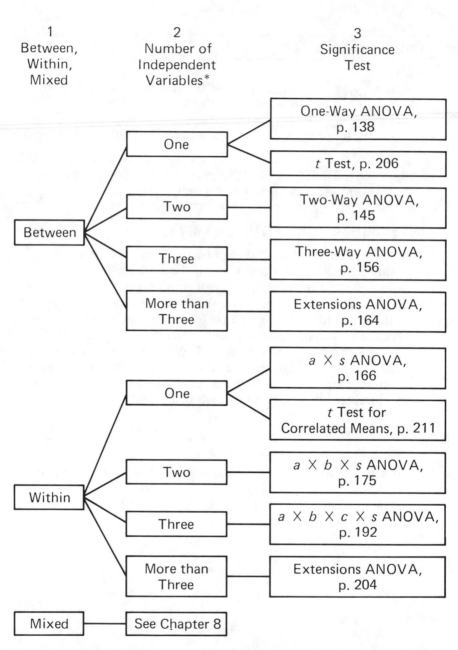

TEST: TEST OF STATISTICAL SIGNIFICANCE
DATA: SCORE

1
Between,
Within,
Mixed

2
Number of
Independent
Variables*

3
Significance
Test

Between

One
- One-Way ANOVA, p. 138
- t Test, p. 206

Two
- Two-Way ANOVA, p. 145

Three
- Three-Way ANOVA, p. 156

More than Three
- Extensions ANOVA, p. 164

Within

One
- $a \times s$ ANOVA, p. 166
- t Test for Correlated Means, p. 211

Two
- $a \times b \times s$ ANOVA, p. 175

Three
- $a \times b \times c \times s$ ANOVA, p. 192

More than Three
- Extensions ANOVA, p. 204

Mixed — See Chapter 8

*Remember, all analyses on this page can be used with any number of levels of each independent variable, except for the t-test, which can be used only when there are two levels.

CHAPTER SEVEN

SCORE DATA

LIMITATIONS AND EXCEPTIONS

The branching program makes three assumptions about the research design. These analyses are appropriate only if each of these assumptions is true:

1. *Fixed-effect model.* The design is a fixed-effects model for analysis of variance. This means that the levels of the independent variable(s) have been arbitrarily chosen by the researcher and that no generalizations are to be made beyond the levels that are studied. The distinction between fixed- and random-effects models is spelled out on p. 128.

2. *Factorial combination.* All designs employing more than one independent variable have factorially combined the several independent variables.

3. *Equal number of scores.* There are an equal number of scores in each treatment or treatment combination. (Exception: The one-way ANOVA and the *t* test may have unequal numbers.) If you have an unequal number of scores, adopt one of the solutions suggested in the section *Unequal number of subjects,* p. 130.

In their coverage of score data, statistics books traditionally teach the *t* test for designs involving two treatment conditions (two levels of the independent variable) and the analysis of variance for more complex designs. We have included a discussion of the *t* test because many instructors feel that it should be available to their students. However, for the following reasons, we recommend that the beginner not complicate his task of mastering score-data analysis by learning the *t* test.

First, the *t* test is merely a special case of the analysis of variance; in fact, the value of t^2 equals F when there are only two treatment conditions. Second, the computing rules for the analysis of variance are simpler than those for the *t* test. Third, the analysis of variance may be used whenever the *t* test is appropriate, but the *t* test may be employed only with *two* treatment conditions. We believe, therefore, that there is no reason to teach the *t* test for simple designs when the analysis of variance is appropriate for simple designs and necessary for complex designs.

Some readers may be surprised by our assertion that the rules for the analysis of variance are relatively simple. Many people mistakenly believe that the analysis of variance must be a difficult technique since it is applied to extremely complex research designs. In truth, the computing rules for analysis of variance are among the simplest for any statistical test of significance. There are many steps in most analyses of variance, but the computations at each step are relatively simple. Moreover, the same computational steps occur over and over in any design, from the simplest to the most complex (see *The Common Pattern for All Analysis of Variance*, p. 136). In performing an analysis of variance, the most important requirements are that you *compulsively* set up your tables in an orderly manner (we provide detailed instructions) and that you know where you are in the analysis at all times. Orderliness and neatness pay off very quickly and are an agreeable alternative to spending hours trying to find an error.

Analysis of variance is one of the most powerful and flexible statistical tests of significance. It is applicable to simple, two-condition experiments, but it can be expanded to analyze research with any number of independent variables and any number of levels of those variables. With only minor modifications, it can be used to analyze between-subjects, within-subjects, and complex designs. This chapter gives computing formulas and examples for between- and within-subject designs. Chapter 8 covers some simple mixed designs.

How the analysis of variance works. How the analysis of variance works may be understood with a little study. We shall illustrate the mechanics with a simple example.

To begin with, groups of scores (ages, height, and so on) are more or less variable. If we wish to know how variable the scores in a population are, we can estimate the population variability by computing the variability of a sample drawn randomly from the population. Samples from the same population, regardless of size and other factors, should yield very similar estimates. Suppose that you randomly draw a sample of three scores from a normally distributed population of such scores. The scores drawn are 1, 2, and 3. From these three scores, you wish to estimate the variance in the parent population. An estimate of population variance ($\hat{\sigma}^2$), derived from a set of sample scores, can be obtained by the use of the formula

$$\hat{\sigma}^2 = \frac{\Sigma X^2 - \frac{(\Sigma X)^2}{N}}{N-1}$$

where X = any score, and

N = the number of scores.

For our sample, $\Sigma X = 6$, $\Sigma X^2 = 14$, and $N = 3$. Substituting these scores into the formula, we obtain

$$\hat{\sigma}^2 = \frac{14 - \frac{6^2}{3}}{2} = \frac{14 - \frac{36}{3}}{2} = \frac{14 - 12}{2} = \frac{2}{2} = 1.$$

Therefore our best estimate of the population variance is that it equals 1. Now let's assume that we draw at random from the population three more scores: 3, 4, and 5. We can now make a separate estimate of the population variance based on these scores. For these scores, $\Sigma X = 12$, $\Sigma X^2 = 50$, and $N = 3$. Substituting these values into the same formula, we obtain

$$\hat{\sigma}^2 = \frac{50 - \frac{(12)^2}{3}}{2} = \frac{50 - \frac{144}{3}}{2} = \frac{50 - 48}{2} = \frac{2}{2} = 1.$$

Thus our best estimate from the second set of scores is also that the population variance equals 1. Let us now assume that we draw at random three more scores: 5, 6, and 7. The estimate of the population variance computed in exactly the same way for these three scores also equals 1. Since we have three estimates, our best overall estimate would be the mean of the three estimates, which would yield us an average estimate of $(1 + 1 + 1)/3 = 1$.

There is, however, *another independent and perfectly legitimate method of estimating the variance of the parent population.* We have three samples and each sample has a mean. The mean of our first sample is 2, of our second is 4, and of our third is 6. These three means can be considered a random sample of the distribution of sample means of sample size 3 for the population in question. If we could estimate the variance of the distribution of sample means, we could estimate the variance of the population from that value. The first step is to estimate the variance of the distribution of sample means $(\hat{\sigma}_{\bar{X}}^2)$, by using a formula very similar to the one we have been using:

$$\hat{\sigma}_{\bar{X}}^2 = \frac{\Sigma \bar{X}^2 - \dfrac{(\Sigma \bar{X})^2}{n}}{n - 1}$$

where $\bar{X}$ = any mean, and

   n = the number of means.

For our three samples, $\Sigma \bar{X} = 12$, $\Sigma \bar{X}^2 = 56$, and $n = 3$. Substituting these values, we obtain

$$\hat{\sigma}_{\bar{X}}^2 = \frac{56 - \dfrac{(12)^2}{3}}{2} = \frac{56 - \dfrac{144}{3}}{2} = \frac{56 - 48}{2} = \frac{8}{2} = 4.$$

Our estimate of the variance of the distribution of sample means is 4. We may now estimate the variance of the population by using the formula $\hat{\sigma}^2 = \hat{\sigma}_{\bar{X}}^2 n$. Substituting in the latter, we obtain $\hat{\sigma}^2 = 4(3) = 12$.

We now have a problem. Using the average variance in each sample to estimate our population variance, we find that the population variance equals 1. Using the sample means to estimate the variance of the distribution of sample means and estimating the population variance from that, we find that the population variance equals 12. These estimates are independent of each other, and each is legitimate. How can we reconcile the fact that one correct estimate is 12 times as large as another correct estimate? The most reasonable explanation is that the three samples were not drawn from the same population but from different populations, having different means but the same variance. Had our three samples been drawn from the same population, two correct and independent methods of estimating population variance would yield the same estimates (except for chance variation).

In this example, we have computed a simple analysis of variance. All analyses of variance make an estimate of the variance in the population by averaging the variance within each condition. (This estimate is called the mean square (*MS*) error.) This is what we did in the first steps above. A second variance estimate uses the means of each condition to estimate the variance of the distribution of sample means. Then an estimate of the population variance (mean-square treatment) is determined from that. The effects of the independent variable are evaluated by computing these two estimates. The ratio of the two estimates (MS_{Error} is always the denominator) yields F. (To see this clearly, perform a one-way analysis of variance using these same sample values, 1, 2, 3; 3, 4, 5; 5, 6, 7. You should obtain an F of 12.)

The analysis of variance tests the null hypothesis that two or more groups have been drawn from the same population of scores. If there are no differences between groups (the null hypothesis is true), the F ratio should be about 1, since the two independent estimates of population variance should be approximately equal. As the differences between the groups become larger (the null hypothesis is false), the estimate based on the variability among means becomes increasingly larger than the estimate averaging the variance within each group. As a consequence, the F ratio becomes larger and larger. When the F ratio becomes large, it is no longer reasonable to believe that the samples came from the same population, and the null hypothesis is rejected.

STATEMENT OF THE NULL HYPOTHESIS

Each analysis of variance tests one or more null hypotheses. In each of the following analyses, there is ordinarily one null hypothesis for each independent variable and a null hypothesis for each of the interactions among independent variables. Each of these kinds of null hypotheses follows the same general form. Because of the great similarity between typical null hypotheses and because of the amount of space required for complete presentation, null hypotheses and specific statistical conclusions are not presented with each example. A general form of each kind of hypothesis and a number of specific examples are presented in this section. If you have difficulty setting up the appropriate null hypothesis when you perform your computations, return to this section and examine these examples again.

MAIN EFFECTS (OF *A*)

General form. The sample means for A (when averaged across all levels of B) were drawn from populations having the same means.
$H_0 : \mu_{A_1} = \mu_{A_2} = \ldots \mu_{A_a}.$

Specific form. The following specific null hypotheses are taken from the examples accompanying the computations in the following pages.
   a. Mean test performances for all test formats were drawn from populations having the same means.
   b. Mean test performances for boys and girls were drawn from populations having the same means.
   c. Mean shot-put performances for each of the three kinds of rewards employed were drawn from populations having the same means.

INTERACTIONS

General form (two-way and three-way). The samples were drawn from populations in which the difference between any two A-level means is the same for each level of B;.similarly, the difference between any two B-level means is the same for each level of A. The samples were drawn from populations in which the interactions between A and B are the same for each level of C; similarly, the interactions between A and C are the same for each level of B and the interactions between B and C are the same for each level of A.

Specific form (two-way). The samples were drawn from populations in which the differences between any two test formats are the same for boys and girls, and, similarly, the differences between boys' and girls' test-performance means are the same for each test format.

STATISTICAL CONCLUSIONS

Each time you obtain an F, you must make a decision with respect to the appropriate null hypothesis that the F tests. In general, if the F is statistically significant (the obtained value is greater than the tabled value of F for the appropriate degrees of freedom), you reject the null hypothesis. If the F is not statistically significant, you do not reject the null hypothesis. Stated formally, the decision about the null hypothesis follows this format:

Main effects (of A). Significant. If the sample means for A (when averaged across all levels of B) were drawn from populations with the same mean, the probability of obtaining means as disparate as the ones obtained in the sample would be less than 5%; therefore the null hypothesis must be rejected.

Main effects (of A). Nonsignificant. If the sample means for A (when averaged across all levels of B) were drawn from populations with the same mean, the probability of obtaining means as disparate as the ones obtained in the sample would *not* be less than 5%; therefore the null hypothesis should *not* be rejected.

A X B interaction. Significant. If the sample means were drawn from populations in which the differences between any two A-level means were the same for each level of B (or the differences between any two B-level means were the same for each level of A), the probability of obtaining differences as discrepant as the ones obtained in the present sample would be less than 5%; therefore the null hypothesis must be rejected.

A X *B interaction. Nonsignificant.* If the sample means were drawn from populations in which the differences between any two *A*-level means were the same for each level of *B* (and the converse), the probability of obtaining differences as discrepant as the ones obtained in the present sample would *not* be less than 5%. Therefore the null hypothesis should *not* be rejected.

VIOLATIONS OF ASSUMPTIONS

Despite the simplicity and flexibility of the analysis of variance, certain conditions must be met before the computing formulas set forth in this chapter can be used. In order to appropriately apply analysis of variance to a set of data, three conditions must be fulfilled: (1) the scores must be from a genuine interval scale,[1] (2) the scores must be normally distributed in the population, and (3) the variance in the treatment conditions or groups must be homogeneous.

However, many statisticians recommend analysis of variance even when the data are not from an interval scale. Other statisticians still challenge this usage, and conservative practice limits the use of analysis of variance to interval data.

There is some good empirical work on the effects of violating the assumptions of normality and homogeneity of variance. In most cases, violations of these assumptions, even fairly extreme ones, do not severely affect the outcome of the analysis of variance. At most, they tend to give a slightly erroneous significance level. For example, although the tabled value may be .05, the actual significance level may range from .07 to .09. Although tests have been developed to determine non-normalcy and heterogeneity of variance, we do not recommend their use. Many of them are less robust than the analysis of variance—many are themselves more susceptible to distortion than is the analysis of variance itself—and most of them are tedious and time-consuming.

---

[1]Since we have attempted to avoid the controversies surrounding problems of scaling and measurement, we refer to data as score, ordered, and frequency rather than the more traditional interval, ordinal, and nominal. For interval data, scale intervals between scores are equal at any point on the scale, and equal intervals between scores are presumed to reflect equal differences in the behaviors being measured. There is considerable debate concerning what scales in the behavioral sciences are in fact genuine interval scales and whether analysis of variance is in fact appropriate when the scores cannot be shown to come from an interval scale. The reader is referred to Kirk (1972, pp. 47–80) for a thorough discussion of this problem and related considerations.

FIXED VS RANDOM MODELS OF THE ANALYSIS OF VARIANCE

In most research designs, the experimenter chooses the levels of his independent variable on the basis of some arbitrary criterion. For example, he may wish to explore the effects of cooperative, individualistic, and competitive instructions on decision-making behavior, or the effect on subsequent learning of written vs oral presentation of verbal material. In both cases, the researcher wishes to see how subjects perform under particular conditions, and he chooses the levels accordingly. He wishes to generalize from the sample results to populations of subjects who would be treated in the same fashion. Generalizing results to populations of subjects who have undergone *specifically chosen levels of an independent variable* or variables defines the fixed-effects model of the analysis of variance. It must be emphasized that it is not safe to generalize beyond the specifically chosen levels in the fixed-effect model.

In other cases, the researcher may randomly select the levels of the independent variable. A population of levels of the independent variable may be available to him and he may randomly choose a certain number of them for his research. For example, he could randomly select ten small towns in the United States with populations under 10,000 to test a new drug, or he could randomly select five levels of food deprivation to study maze learning in the rat. In these cases, the researcher wishes to *generalize the results to a population of levels of the independent variable as well as to a population of subjects*. This type of situation defines the random-effects model of the analysis of variance. In other words, in the random-effects model, the researcher may generalize beyond the subjects *and* beyond the levels of the independent variable that he has actually studied. In the fixed-effects model, he may generalize beyond the subjects but *not beyond the levels* that have been actually tested.

All analyses of variance presented in this book assume a fixed-effects model. These analyses may give misleading results when applied to the random-effects model. The random-effects model represents a considerably more complex situation and may require correction terms or different error terms than are appropriate for the fixed-effects model. If your research design employs a random-effects model or a combination of random and fixed effects, see Kirk (1968), pp. 57, 208–212.

NONFACTORIAL DESIGNS

All analyses of variance in this book appropriate for designs with more than one independent variable assume that the independent variables are factorially combined. Factorial combination, as we have indicated, is the simplest way of studying more than one independent variable in a single study. Problems with recognizing or interpreting nonfactorial combinations may arise in three ways:

1. *A single level of one variable.* A single level of one variable may be studied in combination with (for example) two levels of a second factor. Statistically, this combination presents no problem, since you have only one independent variable. You would have a one-way between-subjects or an *a* X *s* design. For example, you might look at learning under two conditions *for men only*. Your generalizations are restricted to men only, but your analysis is not affected. This design is perfectly appropriate if you are, in fact, interested only in the performance of men.

2. *Confounded variables.* Confounding is a procedure whereby treatments are assigned to subjects so that certain effects cannot be distinguished from other effects. Two factors may be *completely confounded*, as in the example in which men are trained under Learning Condition 1 and women under Learning Condition 2. Although this design may appear to have two factors, gender and training condition, these two factors are completely confounded and their effects cannot be distinguished. If there is a difference between groups, you cannot determine whether the differences are due to gender or training conditions, since the factors covary. Two factors may be *partially confounded*, for example, if subjects trained under Learning Condition 1 were on the average older than subjects trained under Learning Condition 2. In this case, age and training condition are (partially) confounded. Computationally confounded variables present no problem: these studies, for example, may be analyzed with one-way between-subjects analyses of variance; but confounded variables do present major problems of interpretation, since you can never know whether any obtained effects are due to differences in training procedures or to differences in gender or age (there is no statistical way of determining which of the confounded variables produced the effect). Ordinarily, confounding is undesirable and unnecessary and should be avoided because it makes clear interpretation of the outcomes impossible.

3. *Nested designs.* In a nested design, one factor is combined with a second so that some levels of the first factor occur in combination with only one level of the second factor, while other levels of the first factor occur in combination only with the second level of the second factor. The result is a complex design. Although these designs are discussed more fully in Chapter 8, along with relevant sources for additional information, none of these designs are covered in this book. Unlike the first two kinds of nonfactorial designs, the complex design requires a special statistical analysis but does not, when appropriately analyzed, present any special problems of interpretation.

UNEQUAL NUMBER OF SUBJECTS

With one exception, all analyses of variance (as presented in this book) require an equal number of subjects in each treatment combination. The one exception, the one-way between-subjects analysis of variance, does not require equal ns. Although in some situations the researcher may choose to have unequal numbers of subjects (Ss) in his groups, inequality ordinarily occurs as the result of loss of data during the course of the research. In the sections below we discuss, first, the difficulties raised by nonrandom loss of data (along with guidelines for telling when nonrandom loss may have occurred) and then discuss methods for handling the unequal n situation.

The problem of random data loss. One caution must be emphasized in dealing with missing data. Any of these suggested solutions will yield reasonable results *providing* the loss of scores is random—that is, providing the loss is in no way related to the research variable. You should always check your data to be sure you can defend the assertion that the loss is random. Random loss, for example, might involve losing one or more subjects from three of the four cells of a 2 X 2 factorial design. Nonrandom loss of subjects would be suggested if you had lost three subjects from a single cell— perhaps the most difficult cell, the one with the most stringent deprivation level, or the one in which subjects had to return for a session at a later time. In any of these cases, you should consider carefully whether your experimental interpretation may be affected by the loss of subjects. Let us take several examples and examine them more carefully. (1) Suppose you establish a criterion of learning and include only those Ss meeting this criterion. It is possible that you may lose all of the poor learners from the most difficult condition. Ordinarily, the loss of poor learners would make the difficult condition "look better" than it would had a random sample of learners been included. (2) Ordinarily, apparatus failure is a random event; however, if there is any possibility that apparatus failure may be greater for some subjects or conditions (some frustrated animals eat the apparatus), then the loss of Ss may not be random. (3) If Ss must return at a later time, there is the possibility that only "cooperative" Ss will return at the specified time. In a learning experiment, cooperation may not be closely related to learning ability and may not provide a problem for interpretation. On the other hand, loss of cooperative Ss may present a critical problem for a social psychology study in which cooperation was a factor or was related to the factor under study. However, even in the learning study, the possibility that there is poorer cooperation in a difficult condition than in easy conditions should be examined, since differential loss from a difficult condition suggests that the loss is nonrandom.

There are four solutions to the problem of random loss resulting in unequal numbers of Ss: discarding scores, estimating missing scores, using proportional numbers, and using the unweighted means analysis. The first two are quite simple and are usually preferred when they are possible. A combination of the first three methods also may be employed. Ss may be *both* added and dropped to obtain equal n, and Ss may be *both* added and dropped to obtain proportional n.

1. *Randomly discarding scores.* Since in most designs Ss are randomly assigned to conditions, they may also be randomly discarded from conditions. If in a 2 × 2 design you find you have 28, 30, 29, and 32 scores in the four cells, you may randomly discard two scores from the cell that has 30, one score from the cell with 29, and four scores from the cell that has 32. That would leave 28 scores per cell. Notice that we say discard *randomly*. Randomness must be rigidly adhered to. If the Ss or scores are discarded on other than a random basis—for example, if you don't like these scores—you are introducing a serious source of bias into your study. Randomly discarding scores is the simplest technique and is feasible whenever your smallest cell number is reasonably large (say 10).

2. *Estimating missing scores.* An estimate of the missing observations (most often, the mean of the cell from which the score is missing) may be used to replace missing scores. Adding mean scores does not change the estimate of your experimental effect, and, if n is large (at least 10), it changes within-cell variance only slightly. Between 10 and 25% of your scores can be estimated without seriously affecting the analysis of variance. Estimating missing scores is the preferred method when you have small samples, provided that the discrepancies in cell size are not too great (in other words, provided that you need not estimate too large a percentage of the total scores). Remember that your degrees of freedom are based on the number of original observations. The appropriate way to determine the degrees of freedom is to consider the total number of scores (observations and estimated missing

scores) and subtract 1 df from both df_{Error} and df_{Total} for each score that you have added. Both df_{Error} and df_{Total} should be checked carefully. (For complex designs, other dfs including $df_{Subjects}$ may be affected.)

Again considering a 2 × 2 design with 28, 30, 29, and 32 scores in the cells, to use estimated scores you add four scores to the cell with 28 scores, two to the cell with 30 scores, three to the cell with 29 scores, and (of course) none to the group that already has 32 Ss. You have nine estimated scores in all. In each case, the estimated scores that are added are the mean score of the cell to which they are added. You run the analysis of variance as though you had 32 scores in each cell. However, your df_{Total} and df_{Error} would be nine fewer than if you had 32 original observations in each cell. In a 2 × 2 experiment with 32 scores per cell, df_{Total} is 127 and df_{Error} is 124. In this case, they are reduced to 118 and 115, respectively.

3. *Proportional number in cells.* If the number of scores in cells is proportional (for example, if you have 10 scores in A_1B_1 and A_1B_2 and 20 scores in A_2B_1 and A_2B_2), you may perform any of the analyses described by using a slightly modified formula that simply reverses the order of the steps for any sum of squares described in the detailed ANOVA procedures. Instead of following the order:
　　1. Find $\Sigma T_A^2/n_A$.
　　　　a. Square each of the values . . .
　　　　b. Sum the squared values.
　　　　c. Divide the obtained value by the number of scores . . .
Follow this order:
　　2. Find $\Sigma(T_A^2/n_A)$.
　　　　a. Square each of the values from . . .
　　　　b. *Divide each of these squared values by the number of scores on which it is based.*
　　　　c. Sum the obtained values.
Note that in 2b the values differ from case to case but that they must always be proportional. We did not provide the more general procedures for proportional n for each of the analyses because the computational savings from the order followed in number 1 seemed to offset the advantages of providing the general procedure provided by number 2, which is infrequently used.

4. *Unweighted means analysis.* A final alternative, which is somewhat more complicated, is to perform an unweighted means analysis. If there is a random loss of scores from the cells of a design that calls for an equal number of observations, an unweighted means analysis may be performed. This method treats each cell as if it contained the

same number of observations as all other cells. Since a description of this technique is beyond the scope of this book, the reader is referred to Winer (1962), pp. 222, 241, and 374.

A NOTE ON NOTATION

There is no consistent computing-formula notation for analysis of variance in common use among statisticians and researchers. Consequently, we have developed a simplified notational scheme that has more intuitive meaning than most now in use. Although the notational scheme will become clear to you as you go through the programmed examples, we summarize it here.

1. Each independent variable (factor) is labeled with a capital letter: A, B, C, and so on.
2. The number of levels for each independent variable is indicated by a, b, c, and so on. Thus Factor A has a levels. If Factor A were gender, a would be two. Factor B has b levels, Factor C has c levels, and so on.
3. In within-subjects designs, S stands for subjects and s stands for the number of subjects. n_S is the number of measurements on each subject (never the number of subjects).
4. G stands for grand total and refers to the sum of all of the scores in the entire analysis.
5. N always indicates the total *number* of scores in the entire analysis. Thus G is found by adding up all N scores.
6. T stands for total and refers to the sum of the scores for a particular condition. Thus T_{A_1} means the sum of the scores for level A_1 of independent variable A.
7. n refers to the *number* of scores in a particular condition. There are n_{A_1} scores in condition A_1. In other words, T_{A_1} is obtained by adding up n_{A_1} scores. Since all of the analyses except for the one-way analysis of variance require equal ns, we have dropped the numerical subscripts (except in the case of the one-way) for the sake of simplicity, since all ns with the same letter subscript are equal by definition. Thus n_{A_1} always equals n_{A_2}, which always equals n_{A_3}, and so on, and each value is simply called n_A.
8. ΣX^2 always has exactly the same meaning. It is a shorthand expression for the value $\sum_{A=1}^{a} X_A^2$ or $\sum_{A=1}^{a} \sum_{B=1}^{b} X_{AB}^2$, and so on, and it always means: take each of your original scores, square each value, and sum the squared values.

To be sure that you understand each of these notations, we present an example that exemplifies each usage. Refer to Table 7.1.

Table 7.1

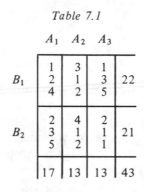

A_1 A_2 A_3

	A_1	A_2	A_3	
B_1	1 2 4	3 1 2	1 3 5	22
B_2	2 3 5	4 1 2	2 1 1	21
	17	13	13	43

Let's assume two independent variables in a factorial between-subjects design.

Independent variable A has three levels. Thus a equals 3.
Independent variable B has two levels. Thus b equals 2.
There are 18 scores in the matrix. Thus N equals 18.
The sum of the 18 scores is 43. Thus G equals 43.
The sum of the six scores in level A_1 is 17. Thus T_{A_1} equals 17.
The sum of the six scores in level A_2 is 13. Thus T_{A_2} equals 13.
The sum of the six scores in level A_3 is 13. Thus T_{A_3} equals 13.
There are six scores in A_1, six in A_2, and six in A_3. Thus n_A equals 6.
The sum of the nine scores in level B_1 is 22. Thus T_{B_1} equals 22.
The sum of the nine scores in level B_2 is 21. Thus T_{B_2} equals 21.
There are nine scores in B_1 and nine in B_2. Thus n_B equals 9.
The sum of the three scores in cell $A_1 B_1$ is 7. Thus $T_{A_1 B_1}$ equals 7.
The sum of the three scores in cell $A_1 B_2$ is 10. Thus $T_{A_1 B_2}$ equals 10.
The sum of the three scores in cell $A_2 B_1$ is 6. Thus $T_{A_2 B_1}$ equals 6.
The sum of the three scores in cell $A_2 B_2$ is 7. Thus $T_{A_2 B_2}$ equals 7.
The sum of the three scores in cell $A_3 B_1$ is 9. Thus $T_{A_3 B_1}$ equals 9.
The sum of the three scores in cell $A_3 B_2$ is 4. Thus $T_{A_3 B_2}$ equals 4.
There are three scores in cell $A_1 B_1$, three in $A_1 B_2$, three in $A_2 B_1$, three in $A_2 B_2$, three in $A_3 B_1$, and three in $A_3 B_2$. Thus n_{AB} equals 3.
If each of the 18 scores is squared and then summed, the total is 135. Thus ΣX^2 equals 135.

CHECKS ON COMPUTATIONS

There are a number of procedures that help ensure both that your arithmetic is correct and that you have included all values required for your computations. Many errors can be avoided if these checks are performed routinely at each relevant step.

A. Check routinely to see that your sums across all independent variables equal G, the grand total. You probably compute G by summing across the totals of one of the independent variables. Check the value against the other sums to ensure accuracy. We have performed the checks for A, B, and AB below.

1. $\sum\limits_{A=1}^{a} T_A = G$

$$17 + 13 + 13 = 43 \qquad \text{OK!}$$

2. $\sum\limits_{B=1}^{b} T_B = G$

$$21 + 22 = 43 \qquad \text{OK!}$$

3. $\sum\limits_{AB=1}^{ab} T_{AB} = G$

$$7 + 10 + 6 + 7 + 9 + 4 = 43 \qquad \text{OK!}$$

B. In each major computational step (such as $\sum\limits_{A=1}^{a} T_A^2/n_A$), every raw score must be included once. A check on computations for each such step may be made by examining the number of scores added to obtain the total (for example, n_A) and then multiplying it by the number of such totals (the number of levels of the factor, in this case, a). This product must always equal N. (Unfortunately, these formulas work even if you haven't performed the operations correctly. You must check to make sure that you have, in fact, summed all of the values represented by this multiplication.)

1. $n_A \times a = N$

$$6 \times 3 = 18 \qquad \text{OK!}$$

2. $n_B \times b = N$

$$9 \times 2 = 18 \qquad \text{OK!}$$

3. $n_{AB} \times a \times b = N$

$$3 \times 3 \times 2 = 18 \qquad \text{OK!}$$

THE COMMON PATTERN FOR ALL ANALYSIS OF VARIANCE

There are basically four related sets of computations involved in all analyses of variance. Recognizing their relationship will greatly speed your mastery of analysis of variance.

1. *Computation of the correction factor:* G^2/N. Most steps in the analysis of variance involve subtracting the correction factor, G^2/N. To obtain this value, sum all the scores. This yields G. Square G and divide this value by N. Once you have obtained this value, record it in a convenient place for easy reference.

2. $SS_{Total} = \Sigma X^2 - G^2/N$. For every analysis of variance, from the simplest to the most complex, computation of SS_{Total} involves the same basic steps: (a) Square each raw score (each of your original values) and sum these squared values. This yields ΣX^2. (b) From this value, subtract the correction factor, G^2/N.

3. $SS_A = \sum\limits_{A=1}^{a} T_A^2/n_A - G^2/N$. For every independent variable (main effect), a single set of computations is involved. In the following analyses, this SS will sometimes be called SS_A, SS_B, SS_C, and so forth. The computations are identical regardless of the label. Take a single independent variable: (a) obtain a total for each level of the variable; (b) square each total; (c) sum the squared totals; (d) divide the sum by the number of scores that were added together to obtain each total.[2] (e) From this value, subtract the correction factor, G^2/N.

4. $SS_{AB} = \sum\limits_{AB=1}^{ab} T_{AB}^2/n_{AB} - G^2/N - SS_A - SS_B$. For every interaction between two (or more) variables, a comparable set of computations is involved. In the following analyses, such SS include SS_{AB}, SS_{AC}, SS_{BC}, SS_{ABC}, and so on. For computing all interactions, we recommend developing an interaction table. Then: (a) obtain a total for each cell of your interaction table; (b) square each total; (c) sum the squared totals; (d) divide the sum by the number of scores added to obtain each total. (e) From this value, subtract the correction factor, G^2/N; (f) from this value, subtract the SS for each (appropriate) main effect and for each (appropriate) lower-order interaction. For example, with SS_{AB}, subtract SS_A and SS_B. For SS_{AC}, subtract SS_A and SS_C. For SS_{ABC}, subtract SS_A, SS_B, SS_C, SS_{AB}, SS_{AC}, and SS_{BC}. For a four-way interaction, you would sub-

---

[2] In the case of the one-way ANOVA with *unequal* number of subjects, steps (c) and (d) must be reversed.

tract each of the four main effects, each of the six two-way interactions, and each of the four three-way interactions. For a five-way
interaction, subtract the main effects, the one-way, two-way, three-
way, and four-way interactions, and so on.

NOTE ON THE ANALYSES AND WORKED EXAMPLES

Each analysis is presented with spelled-out step-by-step procedures. There is a worked example for most analyses. The example
begins with a description of the study, indicates the steps through the
branching program, and goes step by step through the computations.

The three-way ANOVA and the three-way-by-subjects ANOVA
have not been presented in complete detail because of their length; in
addition, because of the length and complexity of the computations,
there are no examples for these three-way designs.

Accompanying each analysis is a brief description that lists in
summary form when the analysis may be used, special requirements,
and so on.

Each analysis provides a reference to the chapters on interpretation, specific comparisons, and strength of association. Each example
includes an interpretation; however, we have included neither the
specific-comparison test nor the strength-of-association measures on
these examples.

Description:

One independent
   variable
a levels of the inde-
   pendent variable
Between-subjects
*n*s need *not* be equal
Both specific compari-
   sons and strength-
   of-association tests
   available
Tests for differences
   between (or among)
   the groups

ONE-WAY

BETWEEN-SUBJECTS ANOVA[3]

Follow through the steps below.

• *Have you read the* Limitations and Exceptions *section?*

• *Are all of these assumptions met for your study? If not, do not continue.*

• *Verbal equivalents for all notations are presented in the* Computations *section below (see also* Note on Notation, *p. 133).*

• *If you are experienced with computations for the analysis of variance, go directly to the Summary Table 7.2 and proceed. If you require step-by-step guidance, begin with Section B,* The Example.

• *If your design has more (or fewer) levels than the present example, simply expand (or contract) the design by continuing any particular step until all levels or level combinations have been accounted for.*

A. (See Table 7.2.)

Table 7.2
Summary Table: One-Way Between-Subjects ANOVA

Source	df	SS = Sum of Squares	MS = Mean Square	F
Factor *A*	$a - 1$	$SS_A = \sum_{A=1}^{a} T_A^2/n_A - G^2/N$	$MS_A = SS_A/(a-1)$	MS_A/MS_{Error}
Error	$N - a$	$SS_{Error} = SS_{Total} - SS_A$	$MS_{Error} = SS_{Error}/(N-a)$	
Total	$N - 1$	$SS_{Total} = \sum X^2 - G^2/N$		

[3] This design is also called a "completely randomized design" or a "randomized groups design."

B. DESCRIPTION OF THE STUDY

1. Mr. Smith, the track coach at Marmaduke High School, wished to determine whether type of reward would affect shot-put performance in tenth-grade boys. He randomly assigned the 21 boys in his class to three experimental groups. The first group of seven boys were told that the boy who put the shot farthest would receive two tickets to the college track meet that evening. The second group of seven boys were told that the boy who did best would be permitted to drive Mr. Smith's sports car the next day. The third group of boys were told that the boy who performed best would be given two prepaid tickets to the X-rated movie at the Bijou that evening. Unfortunately, two boys in the second group sprained their ankles while ogling Mr. Smith's car, and one boy in the third group told his mother about the prize and was not permitted to compete—leaving Mr. Smith with 7, 5, and 6 boys in the three groups, respectively. Since the one-way ANOVA does not require equal numbers in each treatment, he continued with the experiment. The scores in the cells represent the number of feet that each boy put the 16-pound shot.
2. *Null Hypothesis.* See p. 125 for a general statement of the null hypothesis.
3. *Steps through the Branching Program for This Example*
   a. *Statistical technique needed:* **test of significance**
   b. *Kind of data:* **score.** Each subject has a score assigned to him.
   c. *Between, within, mixed:* **between.** Each person receives only one score.
   d. *Number of independent variables:* **one.** Type of reward
   e. *Number of levels of the independent variable:* **Three.** The tickets to the track meet, driving the car, the tickets to the movie

C. RECORDING THE DATA

Construct a raw-data table with *a* columns. Record the data for each condition in the appropriate column (see Table 7.3). Our present example is a one-way between-subjects design with three levels of the factor (a three-column table).

D. INITIAL STEPS

1. Begin a Summary Table. Include the column heads exactly as they are indicated in Table 7.2. Include the row heads (the items listed under "Source"), being sure that you use the appropriate

Table 7.3
Raw-Data Table: One-Way Between-Subjects ANOVA

	Incentive	
Track meet Tickets (A_1)	Sports Car (A_2)	X-Rated Movie (A_3)
4	11	17
7	13	21
6	8	18
8	9	12
2	6	9
9		11
5		

$T_{A_1} = 41$	$T_{A_2} = 47$	$T_{A_3} = 88$	$G = 176$
$(n_{A_1} = 7)$	$(n_{A_2} = 5)$	$(n_{A_3} = 6)$	$(N = 18)$

Table 7.4
Summary Table: One-Way Between-Subjects ANOVA

Source	df	SS	MS	F
Incentive (A)	2	251.72	125.86	10.89
Error	15	173.39	11.56	
Total	17	425.11		

name of Factor A as a head as well as the label A. The Summary Table for the example appears in Table 7.4. Complete the table as you proceed through the computations below. Each critical step is marked by the notation: "Record in Summary Table."

2. Compute the degrees of freedom (*df*), using the formulas in the *df* column in Table 7.2, and record them in your Summary Table (see Table 7.7).

   NOTE: a = number of levels of Factor A; N = total number of scores. In this case, $df_A = a - 1 = 3 - 1 = 2$; $df_{Error} = N - a = 18 - 3 = 15$; $df_{Total} = N - 1 = 18 - 1 = 17$. As a check, note that $df_A + df_{Error} = df_{Total}$ (2 + 15 = 17).

3. There are 13 steps in this analysis. Since later computations refer back to steps by number, you will save time if you list the products from each step by number as you proceed. In addition, you can locate and correct errors in your computations more easily if you have recorded your intermediate steps systematically.

E. COMPUTATIONS

SUPERCHECK: No negative sums of squares (*SS*) may be obtained in the ANOVA. If you obtain a negative *SS* at any point, go back and check the preceding steps. *Never* go on if you have a negative *SS*.

1. a. Find T_{A_1}, T_{A_2}, T_{A_3}.

   Sum the scores in Column A_1 (T_{A_1}). Sum the scores in Column A_2 (T_{A_2}). If you have additional levels, continue until you have summed the scores in each column.

$$T_{A_1} = 41, T_{A_2} = 47, T_{A_3} = 88$$

   b. Find n_{A_1}, n_{A_2}, n_{A_3}.

   Count the number of scores in each column. Record this number, in parentheses, with the total at the bottom of each column.

$$n_{A_1} = 7, n_{A_2} = 5, n_{A_3} = 6$$

2. a. Find G.

   Sum all the scores. Record G at the bottom of the raw-data table.

$$G = \Sigma X = 176$$

CHECK: $T_{A_1} + T_{A_2} + T_{A_3} = G$

$$41 + 47 + 88 = 176. \qquad \text{OK!}$$

   b. Find N.

   Count the total number of scores. This is N. Record N, in parentheses, with G.

$$N = 18$$

CHECK: $n_{A_1} + n_{A_2} + n_{A_3} = N$

$$7 + 5 + 6 = 18. \qquad \text{OK!}$$

3. Find G^2/N.
   Square G and divide this value by N.

$$176^2/18 = 30{,}976/18 = 1720.89$$

4. Find ΣX^2.
   Square each raw score (each of your original values) and sum
   these squared values.

$$4^2 + 7^2 + 6^2 + 8^2 \ldots 9^2 + 11^2 = 2146.00$$

5. Find $SS_{Total} = \Sigma X^2 - G^2/N$.
   Subtract: #4 − #3. Record this value in the appropriate cell in
   the Summary Table.

$$2146.00 - 1720.89 = 425.11$$

CHECK: This value must be positive. G^2/N must always be smaller
than or equal to ΣX^2. If G^2/N is larger than ΣX^2, you have
made an error.

$$G^2/N < \Sigma X^2\colon 1720.89 < 2146.00 \qquad \text{OK!}$$

6. Find $\sum\limits_{A=1}^{a} T_A^2/n_A$.
   a. Square each of the values from #1.

$$41^2 = 1681,\ 47^2 \doteq 2209,\ 88^2 = 7744$$

   b. Divide each squared value by the number of scores (n_A) on
      which the total was based.

$$1681/7 = 240.14,\ 2209/5 = 441.80,\ 7744/6 = 1290.67$$

   c. Sum the obtained values.

$$240.14 + 441.80 + 1290.67 = 1972.61$$

7. Find $SS_A = \sum_{A=1}^{a} T_A^2/n_A - G^2/N$.
   Subtract: #6c − #3. Record in Summary Table.

$$1972.61 - 1720.89 = 251.72$$

8. Find $SS_{Error} = SS_{Total} - SS_A$.
   Subtract: #5 − #7. Record in Summary Table.

$$425.11 - 251.72 = 173.39$$

9. Find $MS_A = SS_A/df_A$.
   Divide #7 by $(a - 1)$. Record in Summary Table.

$$251.72/2 = 125.86$$

10. Find $MS_{Error} = SS_{Error}/df_{Error}$.
    Divide #8 by $(N - a)$. Record in Summary Table.

$$173.39/15 = 11.56$$

11. Find $F = MS_A/MS_{Error}$.
    Divide #9 by #10. Record in Summary Table.

$$125.86/11.56 = 10.89$$

12. Enter the F Table, Appendix 3, with $df_1 = (a - 1)$ and $df_2 = (N - a)$. Note the tabled value of F (at this intersection) for $\alpha = .05$.

$$F_{Tabled}, \alpha = .05 \text{ with } df_1 = 2, \text{ and } df_2 = 15 \text{ is } 3.68.$$

13. Compare your computed value of F (#11) with the tabled value of F (#12).
    a. If $F_{Computed} \geq F_{Tabled}$ (if #11 is greater than or equal to #12), then your difference (or differences) are significant. Reject the null hypothesis.

b. If $F_{Computed} < F_{Tabled}$ (if #11 is less than #12), then your difference (or overall differences) are *not* significant. Do not reject the null hypothesis.

10.89 > 3.68. The differences are significant.
Reject the null hypothesis.

F. INTERPRETATION OF THE EXAMPLE

1. *Statistical conclusion.* See p. 126 for a general statement of a statistical conclusion.
2. *Researcher's conclusion.* The three levels of incentive produce different levels of track performance in high school boys.
3. *Journal summary.* The means for the three levels of incentive were: track meet tickets, 5.86; sports car, 9.40; and X-rated movie, 14.67. Analysis of variance performed on these data indicated that there were significant differences among the means, $F(2, 15) = 14.67, p < .05$. [A specific-comparison test must be performed to determine whether specific means differ from each other.]

G. FURTHER PROCEDURES

1. Interpretation: see Chapter 9, *Introduction* and Section C.
2. Specific comparisons: no test is necessary if $a = 2$. For $a > 2$, see Chapter 10, *Introduction* and Section C.
3. Strength-of-association measures: see Chapter 11, *Introduction* and Section C.

Description:

Two independent
  variables
Any number of levels
  of each variable
Between-subjects
ns must be equal
  (see p. 130)
Fixed model only
Both specific compari-
  sons and strength-
  of-association
  tests available
Tests for differences
  among levels of A,
  differences among
  levels of B, and
  interaction between
  A and B

TWO-WAY ($a \times b$)

BETWEEN-SUBJECTS ANOVA[4]

Follow through the steps below.

- *Have you read the* Limitations and Exceptions *section?*

- *Are all of these assumptions met for your study? If not, do not continue.*

- *Verbal equivalents for all notations are presented in the* Computations *section below (see also* Note on Notation, *p. 133).*

- *If you are experienced with computations for the analysis of variance, go directly to Summary Table 7.5 and proceed. If you require step-by-step guidance, begin with Section B,* The Example.

- *If your design has more (or fewer) levels than the present example, simply expand (or contract) the design by continuing any particular step until all levels or level combinations have been accounted for.*

A. (See Table 7.5.)

B. THE EXAMPLE

    1. Mr. Robbins was instructing his fifth-grade class of 18 boys and 18 girls in the techniques of division and the use of decimals.

---

[4] This analysis is also called a "completely randomized factorial design."

Immediately after the discussion, he gave the students a short quiz. He prepared three different types of quizzes, each containing the same numerical problems but presented in a different format. Six boys and six girls (the control group), randomly chosen, were simply given the basic numbers to divide. Another six boys and six girls, chosen at random, were given the same numbers phrased as consumer problems. The final group of six boys and six girls were also given the same numbers, this time phrased as problems of determining batting and pitching averages in baseball. Mr. Robbins recorded the number of correct answers on each of the quizzes in a 2 X 3 matrix.

2. *Null Hypotheses.* See p. 125 for general statements of the null hypotheses.
3. *Steps through the Branching Program for This Example*
   a. *Statistical technique needed:* **test of significance**
   b. *Kind of data:* **score.** The data are the scores on the quizzes.
   c. *Between, within, mixed:* **between.** Each student takes the quiz in only one of its forms.
   d. *Number of independent variables:* **two.** Gender and quiz type
   e. *Number of levels of each independent variable:* **gender has two levels:** male and female. **Quiz type has three levels:** control, consumer, and baseball.

C. RECORDING THE DATA

1. Record the data in an *a* X *b* raw-data table (See Table 7.6), listing the scores for each condition in the appropriate cell. Our present design has two levels of the first factor and three levels of the second factor (2 X 3 table).

D. INITIAL STEPS

1. Begin a Summary Table. Include the column heads exactly as they are indicated in Table 7.5. Include the row heads (the items listed under "Source"), being sure that you use the appropriate *name of the factors* as well as the labels *A, B,* and so on. The Summary Table for the example appears in Table 7.7. Complete the table as you proceed through the computations below. Each critical step is marked by the notation: "Record in Summary Table."
2. Compute the degrees of freedom (*df*), using the formulas in the *df* column of Table 7.5, and record them in your Summary Table (see Table 7.7).

Table 7.5
Summary Table: Two-Way Between-Subjects ANOVA

Source	df	SS = Sum of Squares	MS = Mean Square	F
Factor A	$a - 1$	$SS_A = \sum_{A=1}^{a} T_A^2/n_A - G^2/N$	$MS_A = SS_A/(a - 1)$	MS_A/MS_{Error}
Factor B	$b - 1$	$SS_B = \sum_{B=1}^{b} T_B^2/n_B - G^2/N$	$MS_B = SS_B/(b - 1)$	MS_B/MS_{Error}
$A \times B$	$(a - 1) \cdot (b - 1)$	$SS_{AB} = \sum_{AB=1}^{ab} T_{AB}^2/n_{AB} - G^2/N - SS_A - SS_B$	$MS_{AB} = SS_{AB}/(a - 1)(b - 1)$	$MS_{AB}/MS_{\text{Error}}$
Error	$N - ab$	$SS_{\text{Error}} = SS_{\text{Total}} - SS_A - SS_B - SS_{AB}$	$MS_{\text{Error}} = SS_{\text{Error}}/(N - ab)$	
Total	$N - 1$	$SS_{\text{Total}} = \sum X^2 - G^2/N$		

NOTE: a = the number of levels of Factor A, b = the number of levels of Factor B, N = the total number of scores, and ab = levels of A times levels of B. In this case, $df_A = a - 1 = 3 - 1 = 2$; $df_B = b - 1 = 2 - 1 = 1$; $df_{AB} = (a - 1)(b - 1) = 2 \times 1 = 2$; $df_{Error} = N - ab = 36 - 6 = 30$; $df_{Total} = N - 1 = 35$. As a check, note that $df_A + df_B + df_{AB} + df_{Error} = df_{Total}$ $(2 + 1 + 2 + 30 = 35)$.

Table 7.6
Raw-Data Table: Two-Way Between-Subjects ANOVA (2 × 3)

		Factor A			
		Control (A_1)	Consumer (A_2)	Baseball (A_3)	
Boys (B_1)	6 4 5 2 1 1	$T_{A_1B_1} = 19$ (6)	6 5 5 3 2 0 $T_{A_2B_1} = 21$ (6)	11 10 9 7 6 5 $T_{A_3B_1} = 48$ (6)	$T_{B_1} = 88$ (18)
Girls (B_2)	6 5 5 2 2 1	$T_{A_1B_2} = 21$ (6)	6 6 5 2 2 1 $T_{A_2B_2} = 22$ (6)	6 5 4 2 2 1 $T_{A_3B_2} = 20$ (6)	$T_{B_2} = 63$ (18)
		$T_{A_1} = 40$ (12)	$T_{A_2} = 43$ (12)	$T_{A_3} = 68$ (12)	$G = 151$ (36)

(Factor B / Gender is the row label for the two body rows.)

3. There are 18 steps in this analysis. Since later computations refer back to steps by number, you will save time if you list the products from each step by number as you proceed. In addition, you can locate and correct errors in your computations more easily if you have recorded your intermediate steps systematically.

Table 7.7
Summary Table: Two-Way Between-Subjects ANOVA (2 × 3)

Source	df	SS	MS	F
Test Format (A)	2	39.39	19.70	4.15
Gender (B)	1	17.36	17.36	3.65
Test Format × Gender ($A \times B$)	2	48.39	24.20	5.09
Error	30	142.50	4.75	
Total	35	247.64		

E. COMPUTATIONS

SUPERCHECK: No negative sums of squares (*SS*) may be obtained in the ANOVA. If you obtain a negative *SS* at any point, go back and check the preceding steps. Never go on if you have a negative *SS*.

1. a. Find $T_{A_1B_1}$, $T_{A_1B_2}$, $T_{A_2B_1}$, $T_{A_2B_2}$, $T_{A_3B_1}$, $T_{A_3B_2}$.
   Sum the scores in each of the *AB* cells. The sum in A_1B_1 is $T_{A_1B_1}$, the sum in A_1B_2 is $T_{A_1B_2}$, and so on.

$T_{A_1B_1} = 19$, $T_{A_1B_2} = 21$, $T_{A_2B_1} = 21$, $T_{A_2B_2} = 22$, $T_{A_3B_1}$
$= 48$, $T_{A_3B_2} = 20$ (see the cells of the raw-data table)

   b. Find n_{AB}. [n_{AB} = number of scores per cell]
   Determine the number of scores that enter into each of the cell totals. For convenience, record n_{AB}, in parentheses, in the table below the totals.

$n_{AB} = 6$ [there are 6 scores per cell]

2. a. Find T_{A_1}, T_{A_2}, T_{A_3}.
   Sum the scores in column A_1. This is T_{A_1}. Sum the scores in column A_2. This is T_{A_2}, and so on.

$T_{A_1} = 40$, $T_{A_2} = 43$, $T_{A_3} = 68$ (see column totals)

   b. Find n_A. [$n_A = b \times n_{AB}$]
   Determine the number of scores that enter into each of the column totals. For convenience, record n_A, in parentheses, in the table below the totals for A.

$n_A = 12$ [$2 \times 6 = 12$]

3. a. Find T_{B_1}, T_{B_2}.
   Sum the scores in row B_1. This is T_{B_1}. Sum the scores in row B_2. This is T_{B_2}, and so on.

$$T_{B_1} = 88, \quad T_{B_2} = 63 \text{ (see row totals)}$$

b. Find n_B. [$n_B = a \cdot \times n_{AB}$]
Determine the number of scores that enter into each of the row totals. For convenience, record n_B, in parentheses, in the table below the totals for B.

$$n_B = 18 \ [3 \times 6 - 18]$$

4. a. Find G.
Sum all the scores. Record G at the bottom of the raw-data table.

$$G = 151 \text{ (column totals = row totals)}$$

CHECK: $T_{A_1} + T_{A_2} + T_{A_3} = T_{B_1} + T_{B_2} = G$

$$40 + 43 + 68 = 88 + 63 = 151. \qquad \text{OK!}$$

b. Find N.
Count the total number of scores. This is N. Record N, in parentheses, with G.

$$N = 36$$

CHECK: $N = a \times b \times n_{AB}$

$$2 \times 3 \times 6 = 36. \qquad \text{OK!}$$

5. Find G^2/N.
Square G and divide this value by N.

$$151^2/36 = 22.801/36 = 633.36$$

6. Find ΣX^2.
Square each raw score and sum these squared values.

$$6^2 + 4^2 + 5^2 + \ldots 6^2 + 5^2 + 4^2 + 2^2 + 2^2 + 1^2 = 881$$

7. Find $SS_{Total} = \Sigma X^2 - G^2/N$.
   Subtract: #6 − #5. Record this value in the appropriate cell in the Summary Table.

$$881 - 633.36 = 247.64$$

CHECK: This value must be positive. G^2/N must be smaller than or equal to ΣX^2. If G^2/N is larger than ΣX^2, you have made an error.

$$G^2/N < \Sigma X^2: 633.36 < 881. \qquad \text{OK!}$$

8. Find $\sum_{A=1}^{a} T_A^2/n_A$.
   a. Square each of the values from #2.

$$40^2 + 43^2 + 68^2 =$$

   b. Sum the squared values.

$$8073$$

   c. Divide the obtained value by n_A, the number of scores in the column.

$$8073/12 = 672.75$$

9. Find $SS_A = \sum_{A=1}^{a} T_A^2/n_A - G^2/N$.
   Subtract: #8c − #5. Record in Summary Table.

$$672.75 - 633.36 = 39.39$$

10. Find $\sum_{B=1}^{b} T_B^2/n_B$.
    a. Square each of the values from #3.

$$88^2 + 63^2 =$$

b. Sum the squared values.

$$11,713$$

c. Divide the obtained value by n_B, the number of scores in the row.

$$11,713/18 = 650.72$$

11. Find $SS_B = \sum\limits_{B=1}^{b} T_B^2/n_B - G^2/N$.
   Subtract: #10c − #5. Record in Summary Table.

$$650.72 - 633.36 = 17.36$$

12. Find $\sum\limits_{AB=1}^{ab} T_{AB}^2/n_{AB}$.
   a. Square each of the values from #1.

$$19^2 + 21^2 + 48^2 + 21^2 + 22^2 + 20^2 =$$

b. Sum the squared values.

$$4431$$

c. Divide the obtained value by n_{AB}, the number of scores in each cell.

$$4431/6 = 738.50$$

13. Find $SS_{AB} = \sum\limits_{AB=1}^{ab} T_{AB}^2/n_{AB} - G^2/N - SS_A - SS_B$.
   Subtract: #12c − #5 − #9 − #11. Record in Summary Table.

$$738.50 - 633.36 - 39.39 - 17.36 = 48.39$$

14. Find $SS_{Error} = SS_{Total} - SS_A - SS_B - SS_{AB}$.
   Subtract: #7 − #9 − #11 − #13. Record in Summary Table.

$$247.64 - 39.39 - 17.36 - 48.39 = 142.50$$

15. Find *MS* by dividing *SS* by appropriate (corresponding) *df*. Record in Summary Table.

---

a. $MS_A = \#9/(a - 1) = 39.39/2 = 19.70$. Record in Summary Table.

b. $MS_B = \#11/(b - 1) = 17.36/1 = 17.36$. Record in Summary Table.

c. $MS_{AB} = \#13/(a - 1)(b - 1) = 48.39/2 = 24.20$. Record in Summary Table.

d. $MS_{Error} = \#14/(N - ab) = 142.50/30 = 4.75$. Record in Summary Table.

---

16. Find *F*, using the formulas in the *F* column of the Summary Table—that is, by dividing $MS_{Treatment}$ by MS_{Error}. Record in Summary Table.

---

a. $F_A = \#15a/\#15d = 19.70/4.75 = 4.15$. Record in Summary Table.

b. $F_B = \#15b/\#15d = 17.36/4.75 = 3.65$. Record in Summary Table.

c. $F_{AB} = \#15c/\#15d = 24.20/4.75 = 5.09$. Record in Summary Table.

---

17. Enter the *F* Table, Appendix 3, and note the tabled value of *F* for $\alpha = .05$ for the following intersections:
    a. $df_1 = a - 1$ and $df_2 = N - ab$.

$df_1 = 2$, $df_2 = 30$ Tabled $F = 3.32$

---

b. $df_1 = b - 1$ and $df_2 = N - ab$.

$df_1 = 1$, $df_2 = 30$ Tabled $F = 4.17$

---

c. $df_1 = (a - 1)(b - 1)$ and $df_2 = N - ab$.

$$df_1 = 2, \; df_2 = 30 \qquad \text{Tabled } F = 3.32$$

18. Compare obtained F with appropriate tabled F. If F_{Computed} $\geq F_{\text{Tabled}}$—that is, if #16a is greater than or equal to #17a, #16b is greater than or equal to #17b, and/or #16c is greater than or equal to #17c—then the difference (or differences) are significant. Reject the null hypothesis. If $F_{\text{Computed}} < F_{\text{Tabled}}$—that is, if #16a is less than #17a, #16b is less than #17b, and/or #16c is less than #17c—then the difference (or differences) are *not* significant. Do not reject the null hypothesis.
    a. Compare #16a with #17a.

> 4.15 > 3.32. The differences for test format are significant.
> Reject the null hypothesis.

b. Compare #16b with #17b.

> 3.65 < 4.17. The difference for gender is not significant.
> Do not reject the null hypothesis.

c. Compare #16c with #17c.

> 5.09 > 3.32. The test format × gender interaction is significant.
> Reject the null hypothesis.

F. INTERPRETATION OF THE EXAMPLE

1. *Statistical conclusion.* See p. 126 for a general statement of the statistical conclusion.
2. *Researcher's conclusion.* Performance on the division problems was superior when the problems were presented in a baseball format; however, this difference was produced by boys' performances. Girls performed at about the same level for all tasks.
3. *Journal summary.* The means for the three kinds of test format were control, 3.33; consumer problems, 3.58; baseball problems, 5.67. Analysis of variance performed on these data indicated that there were significant differences among the means, $\dot{F}(2, 30) = 4.15, p < .05$. [Multiple-comparison tests indicate that only the mean for the baseball format differed significantly from the other means.] However, the interaction

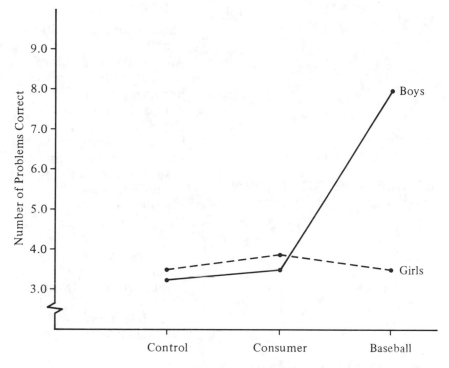

Figure 7.1. *Graphic presentation of the interaction.*

between gender and test format was significant: $F(2, 30) = 5.09, p < .05$. Inspection of Figure 7.1 indicates that superiority of performance on the division problems in the baseball format was due to boys' performances. Girls performed at about the same level for all tasks.

G. FURTHER PROCEDURES

1. Interpretation: see Chapter 9, *Introduction* and Section C for the specific steps through the interpretation of a two-factor design.
2. Specific comparisons: no test is necessary for $a = 2$ and $b = 2$ unless you wish to compare the means of the interaction table. For a or b greater than 2, or if you wish to compare the means of the interaction table, see Chapter 10, *Introduction* and Section C.
3. Strength-of-association measures: see Chapter 11, *Introduction* and Section C.

Description:

Three independent variables

Any number of levels of each variable

Between-subjects

ns must be equal (see p. 130)

THREE-WAY ($a \times b \times c$)

BETWEEN-SUBJECTS ANOVA[5]

Fixed model only

Both specific comparisons and strength-of-association tests available

Tests for differences among levels of A, among levels of B, among levels of C; interactions between AB, AC, and BC; and the interaction among ABC

Follow through the steps below.

- *Have you read the* Limitations and Exceptions *section?*

- *Are all of these assumptions met for your study? If not, do not continue.*

- *Verbal equivalents for all notation are presented in the* Computations *section below (see also* Note on Notation, *p. 133).*

- *If you are experienced with computations for the analysis of variance, go directly to Summary Table 7.8 and proceed. If you require step-by-step guidance, begin with Section B,* Recording the Data.

- *For convenience in simplifying notation, we have used a design with a fixed set of levels. If your design has more (or fewer) levels, simply expand (or contract) the design by continuing any particular step until all levels or level combinations have been accounted for. Because of the length of this analysis, no example is included.*

A. (See Table 7.8.)

---

[5]This analysis is also called a "completely randomized factorial design."

Table 7.8. Summary Table: Three-Way Between-Subjects ANOVA

Source	df	SS = Sum of Squares	MS = Mean Square	F
A	$(a-1)$	$SS_A = \sum\limits_{A=1}^{a} T_A^2/n_A - G^2/N$	$MS_A = SS_A/(a-1)$	$F_A = MS_A/MS_E$
B	$(b-1)$	$SS_B = \sum\limits_{B=1}^{b} T_B^2/n_B - G^2/N$	$MS_B = SS_B/(b-1)$	$F_B = MS_B/MS_E$
C	$(c-1)$	$SS_C = \sum\limits_{C=1}^{c} T_C^2/n_C - G^2/N$	$MS_C = SS_C/(c-1)$	$F_C = MS_C/MS_E$
AB	$\begin{aligned}(a-1)\cdot\\(b-1)\end{aligned}$	$SS_{AB} = \sum\limits_{AB=1}^{ab} T_{AB}^2/n_{AB} - SS_A - SS_B - G^2/N$	$MS_{AB} = SS_{AB}/(a-1)(b-1)$	$F_{AB} = MS_{AB}/MS_E$
AC	$\begin{aligned}(a-1)\cdot\\(c-1)\end{aligned}$	$SS_{AC} = \sum\limits_{AC=1}^{ac} T_{AC}^2/n_{AC} - SS_A - SS_C - G^2/N$	$MS_{AC} = SS_{AC}/(a-1)(c-1)$	$F_{AC} = MS_{AC}/MS_E$
BC	$\begin{aligned}(b-1)\cdot\\(c-1)\end{aligned}$	$SS_{BC} = \sum\limits_{BC=1}^{bc} T_{BC}^2/n_{BC} - SS_B - SS_C - G^2/N$	$MS_{BC} = SS_{BC}/(b-1)(c-1)$	$F_{BC} = MS_{BC}/MS_E$
ABC	$\begin{aligned}(a-1)\cdot\\(b-1)\cdot\\(c-1)\end{aligned}$	$SS_{ABC} = \sum\limits_{ABC=1}^{abc} T_{ABC}^2/n_{ABC} - SS_A - SS_B$ $- SS_C - SS_{AB} - SS_{AC} - SS_{BC}$ $- G^2/N$	$MS_{ABC} = SS_{ABC}/(a-1)(b-1)(c-1)$	$F_{ABC} = MS_{ABC}/MS_E$
Error	$(N-abc)$	$SS_E = SS_{Total} - SS_A - SS_B - SS_C - SS_{AB}$ $- SS_{AC} - SS_{BC} - SS_{ABC}$	$MS_E = SS_E/(N-abc)$	
Total	$N-1$	$SS_{Total} = \sum X^2 - G^2/N$		

B. RECORDING THE DATA

In all the steps below, A, B, and C indicate the names of the three factors; a, b, and c indicate the number of levels of each factor.

1. a. Record the data in an $a \times b \times c$ raw-data table, listing the scores for each condition in the appropriate cell. The A and C factors should be represented in the rows and the B factor in the columns, as shown in Table 7.9. As a convenient case, we shall select for general coverage a three-factor design with two levels of factor A and three levels of both factor B and factor C ($2 \times 3 \times 3$ table).

C. INITIAL STEPS

(These steps are spelled out in greater detail for a somewhat comparable analysis in the *Initial Steps* to the two-way ANOVA, p. 145.)

1. Construct an $a \times b$ subtotal table to contain totals for each of the AB cells. To do this, ignore the third variable (C) and sum together all values for any particular AB combination across the levels of C. Be sure that A is the row factor and B is the column factor, as in the original raw-data table (see Table 7.10a).
2. Construct a $b \times c$ subtotal table to contain totals for each of the BC cells. To do this, ignore the third variable (A) and sum together all values for any particular BC combination across the levels of A. Be sure that C is the row factor and B is the column factor, as in the original raw-data table (see Table 7.10b).
3. Begin a Summary Table. Include the column heads for this table exactly as they are indicated in Table 7.8. Include the row heads (the items listed under "Source"), being sure that you use the appropriate *name of the factors* as well as the labels A, B, and so on. Complete the table as you proceed through the computations below. Each critical step is marked by the notation: "Record in Summary Table."
4. Compute the degrees of freedom (df), using the formulas in the df column of Table 7.8, and record them in your Summary Table.
5. There are 22 steps in this analysis. Since later computations refer back to steps by number, you will save time if you list the products from each step by number as you proceed. In addition, you can locate and correct errors in your computations more easily if you have recorded the intermediate steps systematically.

Table 7.9. Raw-Data Table: Three-Way Between-Subjects ANOVA ($2 \times 3 \times 3$)

		B_1	B_2	B_3	
A_1	C_1	X_1 X_2 $\cdot$ $T_{A_1B_1C_1}$ $\cdot$ (n_{ABC}) $\cdot$ X_n	X_1 X_2 $\cdot$ $T_{A_1B_2C_1}$ $\cdot$ (n_{ABC}) $\cdot$ X_n	X_1 X_2 $\cdot$ $T_{A_1B_3C_1}$ $\cdot$ (n_{ABC}) $\cdot$ X_n	$T_{A_1C_1}$ (n_{AC})
	C_2	X_1 X_2 $\cdot$ $T_{A_1B_1C_2}$ $\cdot$ (n_{ABC}) $\cdot$ X_n	X_1 X_2 $\cdot$ $T_{A_1B_2C_2}$ $\cdot$ (n_{ABC}) $\cdot$ X_n	X_1 X_2 $\cdot$ $T_{A_1B_3C_2}$ $\cdot$ (n_{ABC}) $\cdot$ X_n	$T_{A_1C_2}$ (n_{AC})
	C_3	X_1 X_2 $\cdot$ $T_{A_1B_1C_3}$ $\cdot$ (n_{ABC}) $\cdot$ X_n	X_1 X_2 $\cdot$ $T_{A_1B_2C_3}$ $\cdot$ (n_{ABC}) $\cdot$ X_n	X_1 X_2 $\cdot$ $T_{A_1B_3C_3}$ $\cdot$ (n_{ABC}) $\cdot$ X_n	$T_{A_1C_3}$ (n_{AC})
A_2	C_1	X_1 X_2 $\cdot$ $T_{A_2B_1C_1}$ $\cdot$ (n_{ABC}) $\cdot$ X_n	X_1 X_2 $\cdot$ $T_{A_2B_2C_1}$ $\cdot$ (n_{ABC}) $\cdot$ X_n	X_1 X_2 $\cdot$ $T_{A_2B_3C_1}$ $\cdot$ (n_{ABC}) $\cdot$ X_n	$T_{A_2C_1}$ (n_{AC})
	C_2	X_1 X_2 $T_{A_2B_1C_2}$ $\cdot$ (n_{ABC}) $\cdot$ X_n	X_1 X_2 $T_{A_2B_2C_2}$ $\cdot$ (n_{ABC}) $\cdot$ X_n	X_1 X_2 $T_{A_2B_3C_2}$ $\cdot$ (n_{ABC}) $\cdot$ X_n	$T_{A_2C_2}$ (n_{AC})
	C_3	X_1 X_2 $\cdot$ $T_{A_2B_1C_3}$ $\cdot$ (n_{ABC}) $\cdot$ X_n	X_1 X_2 $\cdot$ $T_{A_2B_2C_3}$ $\cdot$ (n_{ABC}) $\cdot$ X_n	X_1 X_2 $\cdot$ $T_{A_2B_3C_3}$ $\cdot$ (n_{ABC}) $\cdot$ X_n	$T_{A_2C_3}$ (n_{AC})
		T_{B_1} (n_B)	T_{B_2} (n_B)	T_{B_3} (n_B)	G (N)

Table 7.10

a

	B_1	B_2	B_3	
A_1	$T_{A_1B_1}$ (n_{AB})	$T_{A_1B_2}$ (n_{AB})	$T_{A_1B_3}$ (n_{AB})	T_{A_1} (n_A)
A_2	$T_{A_2B_1}$ (n_{AB})	$T_{A_2B_2}$ (n_{AB})	$T_{A_2B_3}$ (n_{AB})	T_{A_2} (n_A)
	T_{B_1} (n_B)	T_{B_2} (n_B)	T_{B_3} (n_B)	G (N)

b

	B_1	B_2	B_3	
C_1	$T_{B_1C_1}$ (n_{BC})	$T_{B_2C_1}$ (n_{BC})	$T_{B_3C_1}$ (n_{BC})	T_{C_1} (n_C)
C_2	$T_{B_1C_2}$ (n_{BC})	$T_{B_2C_2}$ (n_{BC})	$T_{B_3C_2}$ (n_{BC})	T_{C_2} (n_C)
C_3	$T_{B_1C_3}$ (n_{BC})	$T_{B_2C_3}$ (n_{BC})	$T_{B_3C_3}$ (n_{BC})	T_{C_3} (n_C)
	T_{B_1} (n_B)	T_{B_2} (n_B)	T_{B_3} (n_B)	G (N)

D. COMPUTATIONS

SUPERCHECK: No negative sums of squares (SS) may be obtained in the ANOVA. If you obtain a negative SS at any point, go back and check the preceding steps. Never go on if you have obtained a negative SS.

SUPERCHECK: *Each step* from 1 to 7 requires that *every score* be counted (in one category or another) *every time.* Be sure that you systematically count all scores. A check on each of these steps is provided by summing all the totals for any step. *The resulting value must always be G* (see *Checks on Computations*, p. 134).

1. a. Find $T_{A_1B_1C_1}, T_{A_1B_1C_2}, \ldots T_{A_2B_3C_2}, T_{A_2B_3C_3}$.

   Sum the scores in each ABC cell of the $A \times B \times C$ raw-data table. For your convenience, record these values in the appropriate cell of the table. There are $a \times b \times c$ cells.

   b. Find n_{ABC}. [n_{ABC} = number of scores per cell]

   Determine the number of scores on which each of the totals above is based. Record n_{ABC}, in parentheses, in the table immediately below the appropriate totals.

2. a. Find $T_{A_1B_1}, T_{A_1B_2}, T_{A_1B_3}, T_{A_2B_1}, T_{A_2B_2}, T_{A_2B_3}$.

   Sum the scores for each of the six AB combinations—that is, sum all the scores, ignoring C. Record these values in the cells of your $A \times B$ subtotal table.

   b. Find n_{AB}. [$n_{AB} = c \times n_{ABC}$]

   Determine the number of scores on which each of the totals in #2a is based. Record n_{AB}, in parentheses, in the table immediately below the appropriate totals.

3. a. Find $T_{A_1C_1}, T_{A_1C_2}, T_{A_1C_3}, T_{A_2C_1}, T_{A_2C_2}, T_{A_2C_3}$.

   Sum the scores for each of the six AC combinations—that is, sum all the scores, ignoring B. Record these values in the row margins of your $A \times B \times C$ raw-data table.

   b. Find n_{AC}. [$n_{AC} = b \times n_{ABC}$]

   Determine the number of scores on which each of the totals in #3a is based. Record n_{AC}, in parentheses, in the table immediately below the appropriate totals.

4. a. Find $T_{B_1C_1}, T_{B_1C_2}, T_{B_1C_3}, T_{B_2C_1}, T_{B_2C_2}, T_{B_2C_3}, T_{B_3C_1}, T_{B_3C_2}, T_{B_3C_3}$.

   Sum the scores for each of the nine BC combinations—that is, sum all the scores, ignoring A. Record these values in the cells of your $B \times C$ subtotal table.

   b. Find n_{BC}. [$n_{BC} = a \times n_{ABC}$]

   Determine the number of scores on which each of the totals in #4a is based. Record n_{BC}, in parentheses, in the table immediately below the appropriate totals.

5. a. Find T_{A_1}, T_{A_2}.

   Sum the scores in each row of the $A \times B$ subtotal table.

   b. Find n_A. [$n_A = b \times c \times n_{ABC}$]

   Determine the number of scores on which each of the two A-factor totals is based. Record n_A, in parentheses, in the appropriate subtotal table immediately below the totals.

6. a. Find $T_{B_1}, T_{B_2}, T_{B_3}$.

Sum the scores in each column of the $B \times C$ subtotal table.

b. Find n_B. $[n_B = a \times c \times n_{ABC}]$

Determine the number of scores on which each of the three B-factor totals is based. Record n_B, in parentheses, in the appropriate subtotal table immediately below the totals.

7. a. Find $T_{C_1}, T_{C_2}, T_{C_3}$.

Sum the scores in each row of the $B \times C$ subtotal table.

b. Find n_C. $[n_C = a \times b \times n_{ABC}]$

Determine the number of scores on which each of the three C-factor totals is based. Record n_C, in parentheses, in the appropriate subtotal table immediately below the totals.

8. a. Find G.

Sum all the scores. Record G at the bottom of the raw-data table.

CHECK: $T_{A_1} + T_{A_2} = T_{B_1} + T_{B_2} + T_{B_3} = T_{C_1} + T_{C_2} + T_{C_3} = G$

b. Find N.

Count the total number of scores in the $A \times B \times C$ raw-data table. Record N, in parentheses, with G.

CHECK: $N = a \times b \times c \times n_{ABC}$

9. Find G^2/N.

Square G and divide the obtained value by N.

10. Find ΣX^2.

Square each individual's score in the $A \times B \times C$ raw-data table. Then sum the squared values.

11. Find $SS_{\text{Total}} = \Sigma X^2 - G^2/N$.

Subtract: #10 − #9. Record this value in the appropriate cell in the Summary Table.

12. Find $SS_A = \sum_{A=1}^{a} T_A^2/n_A - G^2/N$.

a. Square each of the values from #5a, sum the squared values, and divide by n_A (#5b).

b. From this value subtract #9. Record in Summary Table.

13. Find $SS_B = \sum_{B=1}^{b} T_B^2/n_B - G^2/N$.

a. Square each of the values from #6a, sum the squared values, and divide by n_B (#6b).

b. From this value subtract #9. Record in Summary Table.

14. Find $SS_C = \sum_{C=1}^{c} T_C^2/n_C - G^2/N.$

    a. Square each of the values from #7a, sum the squared values, and divide by n_C (#7b).

    b. From this value subtract #9. Record in Summary Table.

15. Find $SS_{AB} = \sum_{AB=1}^{ab} T_{AB}^2/n_{AB} - SS_A - SS_B - G^2/N.$

    a. Square each of the values from #2a, sum the squared values, and divide by n_{AB} (#2b).

    b. From this value subtract #12b, #13b, and #9. Record in Summary Table.

16. Find $SS_{AC} = \sum_{AC=1}^{ac} T_{AC}^2/n_{AC} - SS_A - SS_C - G^2/N.$

    a. Square each of the values from #3a, sum the squared values, and divide by n_{AC} (#3b).

    b. From this value subtract #12b, #14b, and #9. Record in Summary Table.

17. Find $SS_{BC} = \sum_{BC=1}^{bc} T_{BC}^2/n_{BC} - SS_B - SS_C - G^2/N.$

    a. Square each of the values from #4a, sum the squared values, and divide by n_{BC} (#4b).

    b. From this value subtract #13b, #14b, and #9. Record in Summary Table.

18. Find $SS_{ABC} = \sum_{ABC=1}^{abc} T_{ABC}^2/n_{ABC} \quad SS_A - SS_B - SS_C - SS_{AB}$
    $- SS_{AC} - SS_{BC} - G^2/N.$

    a. Square each of the values from #1a, sum the squared values, and divide by n_{ABC} (#1b).

    b. From this value subtract #12b, #13b, #14b, #15b, #16b, #17b, and #9. Record in Summary Table.

19. Find $SS_{Error} = SS_{Total} - SS_A - SS_B - SS_C - SS_{AB} - SS_{AC} - SS_{BC} - SS_{ABC}.$

    Subtract: #11 − #12b − #13b − #14b − #15b − #16b − #17b − #18b. Record in Summary Table.

20. Compute the eight mean squares indicated in Summary Table 7.8 by dividing each sum of squares by its corresponding *df*. Record in Summary Table.

21. Find the seven *F*s by dividing mean squares by the appropriate error term (mean-square error) as indicated in Summary Table 7.8. Record in Summary Table.

22. Evaluate computed *F*s by consulting the *F* Table, Appendix 3, using the appropriate *df* combination.

E. FURTHER PROCEDURES

1. Interpretation: see Chapter 9, *Introduction* and Section C.
2. Specific-comparison tests: see Chapter 10, *Introduction* and Section C.
3. Strength-of-association measure: see Chapter 11, *Introduction* and Section C.

EXTENSIONS OF THE

BETWEEN-SUBJECTS ANOVA

● *If you are experienced with computations for the analysis of variance, the following recommendations will probably provide a sufficient guideline. If you require step-by-step guidance, return to the three-way ANOVA, p. 156, for more specific instructions.*

● *Reread the section,* The Common Pattern for All Analysis of Variance, *p. 136, before beginning to set up your extension.*

A. DEVELOPING THE SUMMARY TABLE

1. Set up a summary table with the usual headings: *Source, df, SS, MS,* and *F.*
2. a. In listing your sources in a between-subjects table, the general principle is to systematically list your variables (main effects) both by name and by label (*A, B, C,* and so on).
   b. Then list your two-way interactions, three-way interactions, and so on. The two-way interactions comprise all possible two-way combinations of the variables *A, B, C,* and so on. The three-way interactions comprise all possible three-way combinations of the variables, and so on. As a check, we list the number of main effects and interactions for analyses with

increasing numbers of sources of variance. (A glance at Table 7.11 may prove sobering. Any of these extensions means a lot of work.)

Table 7.11

Number of Variables (Main Effects)	Total Number Interactions	2-way	3-way	4-way	5-way	6-way
1	(0)					
2	(1)	1				
3	(4)	3	1			
4	(11)	6	4	1		
5	(26)	10	10	5	1	
6	(57)	15	20	15	6	1

3. Be sure to include the sources, Error, and Total.
4. Degrees of freedom are always computed in exactly the same way. For the main effects, df are always one less than the number of levels for the particular factor. For the interactions, df are always the cross-products of the df for the appropriate main effects. For example, df_{BC} would be $(b - 1)(c - 1)$, while df_{BDE} would be $(b - 1)(d - 1)(e - 1)$. df_{Total} is always $N - 1$, while df_{Error} is always the total number of observations less the cross-product of the number of levels of the main effects. For example, with a four-factor design, df_{Error} would be $N - abcd$; for a five-factor design, df_{Error} would be $N - abcde$, and so on. Note that the sum of all of the dfs in any table sum to df_{Total} or $N - 1$.

B. COMPUTATIONS

For specific computational procedures, read *The Common Pattern for All Analysis of Variance*, p. 136; then, if you require further guidance, follow through the steps (including additional steps at the appropriate places) of the two-way ANOVA or the three-way ANOVA.

Description:

One independent variable
a levels of the variable
Within-subjects
ns must be equal (see
    p. 130)
Both specific comparisons
    and strength-of-association
    tests available
Tests for difference
    between (or among)
    conditions

ONE-WAY ($a \times s$)

WITHIN-SUBJECTS ANOVA[6]

Follow through the steps below.

- *Have you read the* Limitations and Exceptions *section?*

- *Are all of these assumptions met for your study? If not, do not continue.*

- *Verbal equivalents for all notations are presented in the* Computations *section below (see also* Note on Notation, *p. 133).*

- *If you are experienced with computations for the analysis of variance, go directly to Summary Table 7.12 and proceed. If you require step-by-step guidance, begin with Section B,* The Example.

- *If your design has more (or fewer) levels than the present example, simply expand (or contract) the design by continuing any particular step until all levels or level combinations have been accounted for.*

A. (See Table 7.12.)

B. THE EXAMPLE

1. Dr. Jones wished to perform a study on increasing creativity of college-age subjects. Each of 11 randomly selected students was first permitted to solve a set of multiple-solution problems as he ordinarily would. Then each student was asked to watch a movie that detailed the way in which famous inventors find solutions to problems. Dr. Jones believed that the movie would improve problem-solving performance. Following the movie, each student was given an additional set of multiple-solution problems. The data were the average number of solutions to ten problems.

[6] This analysis is also called a "randomized block design."

Table 7.12. Summary Table: One-Way Within-Subjects ANOVA*

Source	df	SS = Sum of Squares	MS = Mean Square	F
Factor A	$a - 1$	$SS_A = \sum_{A=1}^{a} T_A^2/n_A - G^2/N$	$MS_A = SS_A/(a - 1)$	MS_A/MS_{AS}
Subjects (S)	$s - 1$	$SS_S = \sum_{S=1}^{s} T_S^2/n_S - G^2/N$	(not computed)	
A × S	$(a - 1) \cdot (s - 1)$	$SS_{AS} = SS_{Total} - SS_A - SS_S$	$MS_{AS} = SS_{AS}/(a - 1)(s - 1)$	
Total	$N - 1$	$SS_{Total} = \sum X^2 - G^2/N$		

*Most factors are included in a design because we wish to determine whether or not they have an effect on the dependent variable. "Subjects," however, constitutes a somewhat different case. Ordinarily it is assumed that subjects *will* differ from each other; thus this factor is introduced to permit greater control, and the discovery of differences between subjects is of no particular interest. For that reason, the MS_S and F_S are not computed.

2. *Null Hypothesis.* See p. 125 for a general statement of the null hypothesis.
3. *Steps through the Branching Program for This Example*
   a. *Statistical technique needed:* **test of significance**
   b. *Kind of data:* **score.** Number of solutions to ten problems
   c. *Between, within, mixed:* **within.** Each subject is measured both before and after he has seen the movie.
   d. *Number of independent variables:* **one.** Creativity treatment (Note that subjects is NOT ordinarily regarded as an independent variable.)
   e. *Number of levels of independent variable:* **two.** Before and after movie

C. RECORDING THE DATA

1. Record the data in an $a \times s$ table, as shown in Table 7.13. For convenience in following the discussion, make A the column heads and S the row heads. Our present design has two levels of Factor A, with 11 subjects. The number of scores in such a design is 22.

Table 7.13
Raw-Data Table: One-Way Within-Subjects ($2 \times s$)

S Number	Creativity Pretraining (A_1)	Post-training (A_2)	Total
1	4	8	12 (2)
2	3	6	9 (2)
3	5	5	10 (2)
4	4	3	7 (2)
5	6	9	15 (2)
6	2	8	10 (2)
7	4	4	8 (2)
8	3	4	7 (2)
9	2	4	6 (2)
10	5	7	12 (2)
11	3	9	12 (2)
	$T_{A_1} = 41$	$T_{A_2} = 67$	$G = 108$
	(11)	(11)	(22)

D. INITIAL STEPS

1. Begin a Summary Table. Include the column heads exactly as they are indicated in Table 7.12. Include the row heads (the

items listed under "Source"), being sure that you use the appropriate *name of the factor* as well as the labels A, B, and so on. The Summary Table for the example is Table 7.14. Complete the table as you proceed through the computations below. Each critical step is marked by the notation: "Record in Summary Table."

Table 7.14. Summary Table: One-Way Within-Subjects ANOVA

Source	df	SS	MS	F
Tests (A)	1	30.73	30.73	11.26
Subjects (S)	10	37.82	(not computed)	
Tests $\times$ Subjects ($A \times S$)	10	27.27	2.73	
Total	21	95.82		

2. Compute the degrees of freedom (df), using the formulas in the df column of Table 7.12, and record them in your Summary Table (see Table 7.14).

   NOTE: a = number of levels of Factor A, s = number of subjects, N = the total number of scores. In this case, $df_A = a - 1 = 2 - 1 = 1$; $df_S = s - 1 = 11 - 1 = 10$; $df_{AS} = (a - 1)(s - 1) = 1 \times 10 = 10$; $df_{Total} = N - 1 = 22 - 1 = 21$. As a check, note that $df_A + df_S + df_{AS} = df_{Total}$ (1 + 10 + 10 = 21).

3. There are 15 steps in this analysis. Since later computations refer back to steps by number, you will save time if you list the products from each step by number as you proceed. In addition, you can locate and correct errors in your computations more easily if you have recorded your intermediate steps systematically.

E. COMPUTATIONS

SUPERCHECK: No negative sum of squares (SS) may be obtained in the ANOVA. If you obtain a negative SS at any point, go back and check the preceding steps. Never go on if you have a negative SS.

1. a. Find T_{A_1}, T_{A_2}.
   Sum the scores in Column A. This is T_{A_1}. Sum the scores in A_2. This is T_{A_2}. Continue until you have summed the scores in each column.

$$T_{A_1} = 41, \quad T_{A_2} = 67$$

b. Find n_A. $[n_A = s]$
   Count the number of scores in each column. Record this
   number, in parentheses, with the total at the bottom of each
   column. n_A equals the number of subjects.

$$n_A = 11 \text{ [there are 11 subjects]}$$

2. a. Find $T_{S_1}, T_{S_2}, T_{S_3} \ldots$
   Sum all the scores for Subject 1. These comprise the scores
   in the first row. This sum yields T_{S_1}. Then sum all the scores
   for Subject 2. These comprise the scores in the second row.
   This sum yields T_{S_2}. Continue until you have summed the
   scores for each subject.

$$T_{S_1} = 12, \quad T_{S_2} = 9, \quad T_{S_3} = 10, \quad T_{S_4} = 7, \text{ and so on}$$
[see the row totals of the raw-data table]

b. Find n_S. $[n_S = a]$
   Count the number of scores for each subject. This number
   equals n_S. Record this number, in parentheses, with these
   totals.

$$n_S = 2 \text{ [there are two measurements for each subject]}$$

3. a. Find G.
   Sum all the scores. Record G at the bottom of the raw-data
   table.

$$G = 108$$

CHECK: $T_{A_1} + T_{A_2} = T_{S_1} + T_{S_2} + \cdots + T_{S_9} + T_{S_{10}} + T_{S_{11}} = G$

$$41 + 67 = 12 + 9 + 10 + 7 + 15 + 10 + 8 + 7$$
$$+ \ 6 + 12 + 12 = 108. \quad \text{OK!}$$

b. Find N.
Count the total number of scores. This number is N. Record N, in parentheses, with G.

$$N = 22$$

CHECK: $N = n_A \times a$

$$22 = 11 \times 2. \qquad \text{OK!}$$

4. Find G^2/N.
Square G and divide this value by N.

$$108^2/22 = 11{,}664/22 = 530.18$$

5. Find ΣX^2.
Square each raw score and sum the squared values.

$$4^2 + 3^2 + 5^2 + 4^2 \ldots 4^2 + 7^2 + 9^2 = 626$$

6. Find $SS_{\text{Total}} = \Sigma X^2 - G^2/N$
Subtract: #5 − #4. Record this value in the appropriate cell of the Summary Table.

$$626 - 530.18 = 95.82$$

CHECK: This value must be positive. G^2/N must be smaller than or equal to ΣX^2. If G^2/N is larger than ΣX^2, you have made an error.

$$G^2/N < \Sigma X^2 : 530.18 < 626. \qquad \text{OK!}$$

7. Find $\sum_{A=1}^{a} T_A^2/n_A$.
a. Square each of the values from #1.

$$41^2 + 67^2 =$$

b. Sum the squared values.

$$\boxed{6170}$$

c. Divide the obtained value by the number of scores in the column, n_A.

$$\boxed{6170/11 \ = \ 560.91}$$

8. Find $SS_A \ = \ \overset{a}{\underset{A=1}{\Sigma}} T_A^2/n_A \ - \ G^2/N.$
Subtract: #7c $-$ #4. Record in Summary Table.

$$\boxed{560.91 \ - \ 530.18 \ = \ 30.73}$$

9. Find $\overset{s}{\underset{S=1}{\Sigma}} T_S^2/n_S.$
a. Square each of the values from #2.

$$\boxed{12^2 \ + \ 9^2 \ + \ 10^2 \ + \ 7^2 \ldots + \ 12^2 \ + \ 12^2 \ =}$$

b. Sum the squared values.

$$\boxed{1136}$$

c. Divide the obtained value by the number of scores for each subject, n_S.

$$\boxed{1136/2 \ = \ 568.00}$$

10. Find $SS_S \ = \ \overset{s}{\underset{S=1}{\Sigma}} T_S^2/n_S \ - \ G^2/N.$
Subtract: #9c $-$ #4. Record in Summary Table.

$$\boxed{568.00 \ - \ 530.18 \ = \ 37.82}$$

11. Find $SS_{AS} \ = \ SS_{\text{Total}} \ - \ SS_A \ - \ SS_S.$
Subtract: #6 $-$ #8 $-$ #10. Record in Summary Table.

$$\boxed{95.82 - 30.73 - 37.82 = 27.27}$$

12. Find MS, using the formulas in the MS column of the Summary Table—that is, by dividing SS by the appropriate (corresponding) df. Record in Summary Table.

> a. $MS_A = \#8/(a - 1) = 30.73/1 = 30.73$. Record in Summary Table.
> b. $MS_{AS} = \#11/(a - 1)(s - 1) = 27.27/10 = 2.73$. Record in Summary Table.

13. Find $F_A = MS_A/MS_{AS}$.
    Divide #12a by #12b. Record in Summary Table.

$$\boxed{30.73/2.73 = 11.26}$$

14. Enter the F Table, Appendix 3, with $df_1 = (a - 1)$ and $df_2 = (a - 1)(s - 1)$. Note the tabled value of F (at this intersection) for $\alpha = .05$.

> F_{Tabled}, $\alpha = .05$, with $df_1 = 1$, $df_2 = 10$, is 4.96

15. Compare your computed value of F (#13) with the tabled value of F.
    a. If $F_{\text{Computed}} \geq F_{\text{Tabled}}$—that is, if #13 is greater than or equal to #14—then the difference (or differences) are significant. Reject the null hypothesis.
    b. If $F_{\text{Computed}} < F_{\text{Tabled}}$—that is, if #13 is less than #14—then the differences are *not* significant. Do not reject the null hypothesis.

> $11.26 > 4.96$. The differences are significant. Reject the null hypothesis.

F. INTERPRETATION OF THE EXAMPLE

1. *Statistical conclusion.* See p. 126 for a general statement of a statistical conclusion.

2. *Researcher's conclusion.* Exposing college-age students to a model of creative solutions increases their ability to produce good solutions to problems. (There are problems with this interpretation because the treatment is confounded with order.)

3. *Journal summary.* The mean number of solutions for pre- and post-training sessions were 3.72 and 6.09, respectively, $F(1, 10) = 11.26, p < .05$. Exposing college-age students to a model of creative solutions increases their ability to produce good solutions to problems. The order of the treatments was not controlled, and it is possible that these changes are due simply to practice in the task.

G. FURTHER PROCEDURES

1. Interpretation: see Chapter 9, *Introduction* and Section C.
2. Specific comparisons: no test is necessary if $a = 2$. For $a > 2$, see Chapter 10, *Introduction* and Section C.
3. Strength-of-association measures: see Chapter 11, *Introduction* and Section C.

Description:

Two independent variables
Any number of levels of
each variable
Within-subjects
*n*s must be equal
(see p. 130)
Fixed model only
Both specific comparisons
and strength-of-associa-
tion tests available
Tests for differences among
levels of *A*, differences
among levels of *B*, and
interaction between
A and *B*

TWO-WAY (*a* × *b* × *s*)
WITHIN-SUBJECTS ANOVA[7]

Follow through the steps below.

- *Have you read the* Limitations and Exceptions *section?*

- *Are all of these assumptions met for your study? If they are not, do not continue.*

- *Verbal equivalents for all notations are presented in the* Computations *section below (see also* Note on Notation, *p. 133).*

- *If you are experienced with computations for the analysis of variance, go directly to Summary Table 7.15 and proceed in your usual way. If you require step-by-step guidance, begin with Section B*, The Example.

- *If your design has more (or fewer) levels than the present example, simply expand (or contract) the design by continuing any particular step until all levels or level combinations have been accounted for.*

A. (See Table 7.15).

---

[7]This analysis is also called a "randomized block factorial design."

Table 7.15. Summary Table: Two-Way Within-Subjects ANOVA

Source	df	SS = Sum of Squares	MS = Mean Square	F
A	$(a-1)$	$SS_A = \sum_{A=1}^{a} T_A^2/n_A - G^2/N$	$SS_A/(a-1)$	MS_A/MS_{AS}
B	$(b-1)$	$SS_B = \sum_{B=1}^{b} T_B^2/n_B - G^2/N$	$SS_B/(b-1)$	MS_B/MS_{BS}
S	$(s-1)$	$SS_S = \sum_{S=1}^{s} T_S^2/n_S - G^2/N$	(not computed)	
AB	$(a-1) \cdot$ $(b-1)$	$SS_{AB} = \sum_{AB=1}^{ab} T_{AB}^2/n_{AB} - SS_A - SS_B - G^2/N$	$SS_{AB}/(a-1)(b-1)$	MS_{AB}/MS_{ABS}
AS	$(a-1) \cdot$ $(s-1)$	$SS_{AS} = \sum_{AS=1}^{as} T_{AS}^2/n_{AS} - SS_A - SS_S - G^2/N$	$SS_{AS}/(a-1)(s-1)$	
BS	$(b-1) \cdot$ $(s-1)$	$SS_{BS} = \sum_{BS=1}^{bs} T_{BS}^2/n_{BS} - SS_B - SS_S - G^2/N$	$SS_{BS}/(b-1)(s-1)$	
ABS	$(a-1) \cdot$ $(b-1) \cdot$ $(s-1)$	$SS_{ABS} = SS_{Total} - SS_A - SS_B - SS_S - SS_{AB} - SS_{AS} - SS_{BS}$	$SS_{ABS}/(a-1)(b-1)(s-1)$	
Total	$(N-1)$	$SS_{Total} = \sum X^2 - G^2/N$		

B. THE EXAMPLE

1. A group of scientists were interested in state-specific learning effects with phenobarbital. They selected two matched tasks. Each of ten subjects learned one of the tasks under phenobarbital and learned the other task without phenobarbital. Following training, each subject was tested four times: Task 1, which was learned under phenobarbital, was tested both under phenobarbital and without phenobarbital. Task 2, which was learned without phenobarbital, was tested both under phenobarbital and without phenobarbital. The order of the tests was randomly determined for every subject. The data represent the time the subject required to perform the task during the testing. (Note that no data were obtained during training.)

2. *Null hypotheses.* See p. 125 for general statements of the null hypotheses.

3. *Steps through the Branching Program for This Example*
   a. *Statistical technique needed:* **test of significance**
   b. *Kind of data:* **score.** Time required to perform the task during testing
   c. *Between, within, mixed:* **within.** Each subject receives four scores.
   d. *Number of independent variables:* **Two.** Drug during test, drug during training
   e. *Number of levels of each independent variable:* **condition during training has two levels:** drug and no drug. **Condition during test has two levels:** drug and no drug. [Designs of this sort present difficulty for many people apparently because (1) it seems that there should be scores from the original training sessions and there are none, and (2) there are problems in understanding this kind of factorial combination. One of the problems seems to be that task (Task 1 and Task 2) is not an additional independent variable—since it is completely confounded with treatment. That is, Task 1 and learning under phenobarbital occur together, and Task 2 and learning with no drug occur together.]

C. RECORDING THE DATA

1. Record the data in an $a \times b \times s$ table. For convenience in following the discussion, make S the row heads and B the subheads for columns under the major A column heads, as shown in Table 7.16. Our present design has two levels of each factor with ten subjects ($2 \times 2 \times s$).

Table 7.16. Raw-Data Table: Two-Way Within-Subjects (2 × 2 × s)

S Number	Training under Phenobarbital (A_1)		Training without Phenobarbital (A_2)		T_S
	Test under Phenobarbital (B_1)	Test without Phenobarbital (B_2)	Test under Phenobarbital (B_1)	Test without Phenobarbital (B_2)	
1	6	4	3	6	19 (4)
2	7	2	2	3	14 (4)
3	5	3	5	8	21 (4)
4	8	5	4	5	22 (4)
5	3	6	2	5	16 (4)
6	6	4	7	7	24 (4)
7	8	5	3	6	22 (4)
8	9	2	4	6	21 (4)
9	3	4	3	4	14 (4)
10	6	2	3	5	16 (4)
	$T_{A_1B_1} = 61$ (10)	$T_{A_1B_2} = 37$ (10)	$T_{A_2B_1} = 36$ (10)	$T_{A_2B_2} = 55$ (10)	189
	$T_{A_1} = 97$ (20)		$T_{A_2} = 92$ (20)		

D. INITIAL STEPS

1. Construct three additional tables to contain totals for each of the following factor combinations: $A \times B$, $A \times S$, and $B \times S$. For each table, make the levels of the first factor the column heads and the levels of the second factor the row heads. The totals for each factor combination and n, the number of scores on which each total is based, should be entered as they are computed in steps #1, #2, and #3, respectively (see Table 7.17).

Table 7.17. $A \times B$ Table: Two-way Within-Subjects ANOVA

a

Training Condition

		With Phenobarbital (A_1)	Without Phenobarbital (A_2)	
Test Condition	With Phenobarbital (B_1)	$T_{A_1B_1} = 61$ (10)	$T_{A_2B_1} = 36$ (10)	$T_{B_1} = 97$ (20)
	Without Phenobarbital (B_2)	$T_{A_1B_2} = 37$ (10)	$T_{A_2B_2} = 55$ (10)	$T_{B_2} = 92$ (20)
		$T_{A_1} = 98$ (20)	$T_{A_2} = 91$ (20)	$G = 189$ (40)

$A \times S$ Table: Two-Way Within-Subjects ANOVA

b

Training Condition

		With Phenobarbital (A_1)	Without Phenobarbital (A_2)	T_S
Subjects	1	10 (2)	9 (2)	19 (4)
	2	9 (2)	5 (2)	14 (4)
	3	8 (2)	13 (2)	21 (4)
	4	13 (2)	9 (2)	22 (4)
	5	9 (2)	7 (2)	16 (4)
	6	10 (2)	14 (2)	24 (4)
	7	13 (2)	9 (2)	22 (4)
	8	11 (2)	10 (2)	21 (4)
	9	7 (2)	7 (2)	14 (4)
	10	8 (2)	8 (2)	16 (4)

$B \times S$ Table: Two-Way Within-Subjects ANOVA

c

Test Condition

		With Phenobarbital (B_1)	Without Phenobarbital (B_2)	T_S
	1	9 (2)	10 (2)	19 (4)
	2	9 (2)	5 (2)	14 (4)
	3	10 (2)	11 (2)	21 (4)
	4	12 (2)	10 (2)	22 (4)
Subjects	5	5 (2)	11 (2)	16 (4)
	6	13 (2)	11 (2)	24 (4)
	7	11 (2)	11 (2)	22 (4)
	8	13 (2)	8 (2)	21 (4)
	9	6 (2)	8 (2)	14 (4)
	10	9 (2)	7 (2)	16 (4)

2. Begin a Summary Table. Include the column heads exactly as they are indicated in Table 7.15. Include the row heads (the items listed under "Source"), being sure that you use the appropriate *name of the factor* as well as the labels *A*, *B*, and so on. The Summary Table for the example appears in Table 7.18. Complete the table as you proceed through the computations below. Each critical step is marked by the notation: "Record in the Summary Table."

3. Compute the degrees of freedom (*df*), using the formulas in the *df* column of Table 7.15, and record them in your Summary Table (see Table 7.18).

NOTE: a = number of levels of Factor A; b = number of levels of Factor B; s = number of subjects; N = total number of scores. In this case, $df_A = a - 1 = 2 - 1 = 1$; $df_B = b - 1 = 2 - 1 = 1$; $df_S = s - 1 = 10 - 1 = 9$; $df_{AB} = (a - 1)(b - 1) = 1 \times 1 = 1$; $df_{AS} = (a - 1)(s - 1) = 1 \times 9 = 9$; $df_{BS} = (b - 1)(s - 1) = 1 \times 9 = 9$; $df_{ABS} = (a - 1)(b - 1)(s - 1) = 1 \times 1 \times 9 = 9$; $df_{Total} = N - 1 = 40 - 1 = 39$. As a check, note that $df_A + df_B + df_S + df_{AB} + df_{AS} + df_{BS} + df_{ABS} = df_{Total}$ ($1 + 1 + 9 + 1 + 9 + 9 + 9 = 39$).

4. There are 27 steps in this analysis. Since later computations refer back to steps by number, you will save time if you list the

products from each step by number as you proceed. In addition, you can locate and correct errors in your computations more easily if you have recorded intermediate steps systematically.

Table 7.18. Summary Table: Two-Way Within-Subjects (2 × 2 × *s*)

Source	df	SS	MS	F
Training Condition (*A*)	1	1.22	1.22	< 1
Test Condition (*B*)	1	.62	.62	< 1
Subjects (*S*)	9	29.72	(not computed)	
Training × Test (*A* × *B*)	1	46.23	46.23	22.71
Training × Subjects (*A* × *S*)	9	22.53	2.50	
Test × Subjects (*B* × *S*)	9	23.13	2.57	
A × *B* × *S*	9	18.52	2.06	
Total	39	141.97	(not computed)	

D. COMPUTATIONS

SUPERCHECK: No negative sums of squares (*SS*) may be obtained in the ANOVA. If you obtain a negative *SS* at any point, go back and check the preceding steps. Never go on if you have a negative *SS*.

SUPERCHECK: *Each step* from 1 to 7 requires that *every score* be counted (in one category or another) *every time.* Be sure that you systematically count all scores. A check on each of these steps can be made by summing all the totals for any step. *The resulting value must always be G* (see *Checks on Computations*, p. 134).

1. a. Find $T_{A_1B_1}$, $T_{A_1B_2}$, $T_{A_2B_1}$, $T_{A_2B_2}$.
   Sum the scores in Column A_1B_1. This sum is $T_{A_1B_1}$. Sum the scores in Column A_1B_2, Column A_2B_1, and so on, and continue until you have summed the scores in each *AB* column. Record these values in the appropriate cells of the $A \times B$ table, Table 7.17a.

   $$T_{A_1B_1} = 61, \quad T_{A_1B_2} = 37, \quad T_{A_2B_1} = 36, \quad T_{A_2B_2} = 55$$

   b. Find $n_{AB} \cdot [n_{AB} = s]$
   n_{AB}, the number of scores on which each of the *AB* cell totals is based, equals *s*. Record n_{AB}, in parentheses, in the $A \times B$ table below the corresponding totals.

$$n_{AB} = 10 \text{ [there are 10 subjects]}$$

2. a. Find $T_{A_1 S_1}$, $T_{A_2 S_1}$, $T_{A_1 S_2}$, $T_{A_2 S_2}$, and so on.
      For each subject, sum together all scores for A_1 (by summing
      across all levels of B), and then sum together all scores for A_2
      (by summing across all levels of B). Record these totals in the
      cells of your $A \times S$ table (Table 7.17b).

$$T_{A_1 S_1} = 10, \ T_{A_2 S_1} = 9, \ T_{A_1 S_2} = 9, \ T_{A_2 S_2} = 5, \text{ and so on}$$

   b. Find n_{AS}. $[n_{AS} = b]$
      n_{AS}, the number of scores on which each of the AS cell totals
      is based, equals b. For convenience, record n_{AS}, in parentheses,
      in the $A \times S$ table immediately below the corresponding
      totals.

$$n_{AS} = 2 \ [b = 2]$$

3. a. Find $T_{B_1 S_1}$, $T_{B_2 S_1}$, $T_{B_1 S_2}$, $T_{B_2 S_2}$, and so on.
      For each subject, sum together all scores for B_1 (by summing
      across all levels of A), and then sum together all scores of B_2
      (by summing across all levels of A). Record these totals in the
      cells of your $B \times S$ table (Table 7.17c).

$$T_{B_1 S_1} = 9, \ T_{B_2 S_1} = 10, \ T_{B_1 S_2} = 9, \ T_{B_2 S_2} = 5, \text{ and so on}$$

   b. Find n_{BS}. $[n_{BS} = a]$
      n_{BS}, the number of scores on which each of the BS cells totals
      is based, equals a. For convenience, record n_{BS}, in parentheses,
      in the $B \times S$ table immediately below the corresponding totals.

$$n_{BS} = 2 \ [a = 2]$$

4. a. Find T_{S_1}, T_{S_2}, T_{S_3}, T_{S_4}, and so on.
      Sum all the scores for Subject 1. These comprise all the $a \times b$
      scores in the first row ($2 \times 2 = 4$ in the present case). Then
      sum all the scores for Subject 2. These comprise the $a \times b$
      scores in the second row. Continue until you have summed
      the scores for each subject. You will have s sums.

$$T_{S_1} = 19, \ T_{S_2} = 14, \ T_{S_3} = 21, \ T_{S_4} = 22, \ldots T_{S_{10}} = 16$$

b. Find n_S. [$n_S = a \times b$]

n_S, the number of scores on which each total for subjects is based, equals $a \times b$. For convenience, record n_S, in parentheses, below the corresponding totals.

$$n_S = 4 \ [2 \times 2 = 4]$$

5. a. From the $A \times B$ table, find T_{A_1}, T_{A_2}.

Sum the scores in Column A_1. This sum is T_{A_1}. Sum the scores in Column A_2. This sum is T_{A_2}. Continue until you have summed the scores in each column.

$$T_{A_1} = 98, \ T_{A_2} = 91$$

b. Find n_A. [$n_A = b \times s$]

n_A, the number of scores on which each column total is based, equals $b \times s$. Record n_A, in parentheses, below the corresponding totals.

$$n_A = 20 \ [2 \times 10 = 20]$$

6. a. From the $A \times B$ table, find T_{B_1}, T_{B_2}.

Sum the scores in row B_1. This sum is T_{B_1}. Sum the scores in row B_2. This sum is T_{B_2}. Continue until you have summed the scores in each row.

$$T_{B_1} = 97, \ T_{B_2} = 92$$

b. Find n_B. [$n_B = a \times s$]

n_B, the number of scores on which each row total is based, equals $a \times s$. Record n_B, in parentheses, with the corresponding totals.

$$n_B = 20 \ [2 \times 10 = 20]$$

7. a. Find G.

   Sum all the scores. Record G at the bottom of the raw-data table.

$$\boxed{G = 189}$$

CHECK: $T_{A_1} + T_{A_2} = T_{B_1} + T_{B_2} = T_{S_1} + T_{S_2} + T_{S_3} + \ldots T_{S_{10}}$
$= G$

$$\boxed{\begin{array}{l} 98 + 91 = 97 + 92 = 19 + 14 + 21 + 22 + 16 \\ + 24 + 22 + 21 + 14 + 16 = 189. \qquad \text{OK!} \end{array}}$$

   b. Find N.

   Count the total number of scores. This total is N. Record N, in parentheses, with G.

$$\boxed{N = 40}$$

CHECK: $N = a \times b \times n_{AB}$

$$\boxed{2 \times 2 \times 10 = 40. \qquad \text{OK!}}$$

8. Find G^2/N.

   Square G and divide this value by N.

$$\boxed{189^2/40 = 35{,}721/40 = 893.03}$$

9. Find ΣX^2.

   Square each score (in the original raw-score table) and sum the squared values.

$$\boxed{\begin{array}{c} \Sigma X^2 = 6^2 + 7^2 + 5^2 + 8^2 \ldots 4^2 + 2^2 \ldots 3^2 + 2^2 \ldots \\ 6^2 + 3^2 \ldots 4^2 + 5^2 = 1035 \end{array}}$$

10. Find $SS_{\text{Total}} = \Sigma X^2 - G^2/N$.

    Subtract: #9 − #8. Record this value in the appropriate cell in the Summary Table.

$$1035 - 893.03 = 141.97$$

CHECK: This value must be positive. G^2/N must be smaller than or equal to ΣX^2. If G^2/N is larger than ΣX^2, you have made an error.

$$G^2/N < X^2: 893.03 < 1035. \qquad \text{OK!}$$

11. From the $A \times B$ table, find $\sum\limits_{A=1}^{a} T_A^2/n_A$.

   a. Square each of the values from #5.

$$98^2 + 91^2 =$$

   b. Sum the squared values.

$$17{,}885$$

   c. Divide the obtained value by the number of scores in the column, n_A.

$$17{,}885/20 = 894.25$$

12. Find $SS_A = \sum\limits_{A=1}^{a} T_A^2/n_A - G^2/N$.

   Subtract: #11c − #8. Record in Summary Table.

$$894.25 - 893.03 = 1.22$$

13. Find $\sum\limits_{B=1}^{b} T_B^2/n_B$.

   a. Square each of the values from #6.

$$97^2 + 92^2 =$$

   b. Sum the squared values.

$$17{,}873$$

c. Divide the obtained value by the number of scores in the row, n_B.

$$17{,}873/20 = 893.65$$

14. Find $SS_B = \sum_{B=1}^{b} T_B^2/n_B - G^2/N$.

Subtract: #13c − #8. Record in Summary Table.

$$893.65 - 893.03 = .62$$

15. Find $\sum_{S=1}^{s} T_S^2/n_S$.

a. Square each of the values from #4.

$$19^2 + 14^2 + 21^2 + 22^2 \ldots + 16^2 =$$

b. Sum the squared values.

$$3691$$

c. Divide the obtained value by the number of scores for each subject, n_S.

$$3691/4 = 922.75$$

16. Find $SS_S = \sum_{S=1}^{s} T_S^2/n_S - G^2/N$.

Subtract: #15c − #8. Record in Summary Table.

$$922.75 - 893.03 = 29.72$$

17. From the $A \times B$ table, find $\sum_{AB=1}^{ab} T_{AB}^2/n_{AB}$.

Take the totals of the AB cells (these are the values from #1).

a. Square each of the values of the AB cells.

$$61^2 + 36^2 + 37^2 + 55^2 =$$

b. Sum the squared values.

$$\boxed{9411}$$

c. Divide the obtained value by the number of scores in the AB cells, n_{AB}.

$$\boxed{9411/10 = 941.10}$$

18. Find $SS_{AB} = \overset{ab}{\underset{AB=1}{\Sigma}} T^2_{AB}/n_{AB} - G^2/N - SS_A - SS_B$.
   Subtract: #17c − #8 − #12 − #14. Record in Summary Table.

$$\boxed{941.10 - 893.03 - 1.22 - .62 = 46.23}$$

19. From the $A \times S$ table, find $\overset{as}{\underset{AS=1}{\Sigma}} T^2_{AS}/n_{AS}$.
   Take the totals of the AS cells (these are the values from #2).
   a. Square each of the values of the AS cells.

$$\boxed{10^2 + 9^2 + 8^2 + \ldots + 8^2 =}$$

b. Sum the squared values.

$$\boxed{1893}$$

c. Divide the obtained values by the number of scores in the AS cells, n_{AS}.

$$\boxed{1895/2 = 946.50}$$

20. Find $SS_{AS} = \overset{as}{\underset{AS=1}{\Sigma}} T^2_{AS}/n_{AS} - G^2/N - SS_A - SS_S$.
   Subtract: #19c − #8 − #12 − #16. Record in Summary Table.

$$\boxed{946.50 - 893.03 - 1.22 - 29.72 = 22.53}$$

21. From the $B \times S$ table, find $\overset{bs}{\underset{BS=1}{\Sigma}} T^2_{BS}/n_{BS}$.
   Take the totals of the BS cells (these are the values from #3).
   a. Square each of the values of the BS cells.

$$9^2 + 9^2 + 10^2 + \ldots + 7^2 =$$

b. Sum the squared values.

$$1893$$

c. Divide the obtained values by the number of scores in the *BS* cells, n_{BS}.

$$1893/2 = 946.50$$

22. Find $SS_{BS} = \sum_{BS=1}^{bs} T_{BS}^2/n_{BS} - G^2/N - SS_B - SS_S$.
    Subtract: #21c − #8 − #14 − #16. Record in Summary Table.

$$946.50 - 893.03 - .62 - 29.72 = 23.13$$

23. Find $SS_{ABS} = SS_{Total} - SS_A - SS_B - SS_S - SS_{AB} - SS_{AS} - SS_{BS}$.
    Subtract: #10 − #12 − #14 − #16 − #18 − #20 − #22. Record in Summary Table.

$$141.97 - 1.22 - .62 - 29.72 - 46.23 - 22.53 - 23.13 = 18.52$$

24. Find *MS*, using the formulas in the *MS* column of the Summary Table–that is, by dividing *SS* by the appropriate (corresponding) *df*. Record in Summary Table.

a. $MS_A = \#12/(a-1) = 1.22/1 = 1.22$

b. $MS_B = \#14/(b-1) = .62/1 = .62$

c. $MS_{AB} = \#18/(a-1)(b-1) = 46.23/1 = 46.23$

d. $MS_{AS} = \#20/(a-1)(s-1) = 22.53/9 = 2.50$

e. $MS_{BS} = \#22/(b-1)(s-1) = 23.13/9 = 2.57$

f. $MS_{ABS} = \#23/(a-1)(b-1)(s-1) = 18.52/9 = 2.06$

25. Find F, using the formulas in the F column of the Summary Table—that is, by dividing $MS_{\text{Treatment}}$ by MS_{Error}. Record in Summary Table.

> a. $F_A = \#24a/\#24d = 1.22/2.50 < 1$
>
> b. $F_B = \#24b/\#24e = .62/2.57 < 1$
>
> c. $F_{AB} = \#24c/\#24f = 46.23/2.06 = 22.44$

26. Enter the F Table, Appendix 3, and note tabled value of F for $\alpha = .05$ for the following intersections.

a. $df_1 = a - 1$ and $df_2 = (a - 1)(s - 1)$.

b. $df_1 = b - 1$ and $df_2 = (b - 1)(s - 1)$.

c. $df_1 = (a - 1)(b - 1)$ and $df_2 = (a - 1)(b - 1)(s - 1)$.

> a. $df_1 = 1, df_2 = 9$
>
> b. $df_1 = 1, df_2 = 9$ Tabled $F = 4.21, \alpha = .05$
>
> c. $df_1 = 1, df_2 = 9$

27. Compare obtained F with appropriate tabled F. If $F_{\text{Computed}} \geq F_{\text{Tabled}}$—that is, if $\#25a$, $\#25b$, and/or $\#25c$ are greater than or equal to the tabled values from $\#26$—then the difference or differences are significant. Reject the null hypothesis. If $F_{\text{Computed}} < F_{\text{Tabled}}$—that is, if the computed values $\#25a$, $\#25b$, and/or $\#25c$ are less than the tabled value from $\#26$—then the difference or differences are not significant. Do not reject the null hypothesis.

a. Compare $\#25a$ with $\#26a$.

> All values less than 1 are nonsignificant.
> Do not reject the null hypothesis.

b. Compare $\#25b$ with $\#26b$.

> All values less than 1 are nonsignificant.
> Do not reject the null hypothesis.

c. Compare #25c with #26c.

> 22.44 > 4.21. The differences are significant.
> Reject the null hypothesis.

F. INTERPRETATION OF THE EXAMPLE

1. *Statistical conclusion.* See p. 126 for a general statement of the statistical conclusions.
2. *Researcher's conclusion.* Performance of subjects is superior if they are trained and tested with phenobarbital or if they are trained and tested without phenobarbital. Their performance is inferior if they are trained under one condition and tested under the other.

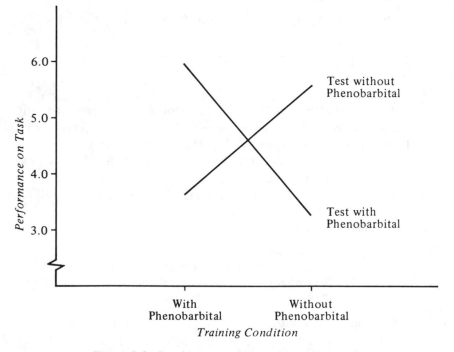

Figure 7.2. *Graphic presentation of the interaction.*

3. *Journal summary*. There were no significant main effects indicating that the condition under which training or test occurred did not, by itself, differentially influence performance (see Figure 7.2). There was, however, a significant interaction between training and test conditions, $F(1, 9) = 22.44, p < .05$, indicating that performance on the test is specific to the state in which learning occurred. Subjects performed well on the task when they were trained and tested with phenobarbital or when they were trained and tested without phenobarbital. Their performance was inferior if they were trained under one condition and tested under another.

G. FURTHER PROCEDURES

1. Interpretation: see Chapter 9, *Introduction* and Section C.
2. Specific comparisons: see Chapter 10, *Introduction* and Section C.
3. Strength-of-association measures: see Chapter 11, *Introduction* and Section C.

Description:

Three independent
  variables
Any number of levels of
  each variable
Within-subjects
ns must be equal (see
  p. 130)
Fixed model only
Both specific comparisons

THREE-WAY ($a \times b \times c \times s$)

WITHIN-SUBJECTS ANOVA[8]

  and strength-of-associ-
  ation tests available
Tests for differences among
  levels of A, among levels
  of B, among levels of C;
  interactions between AB,
  AC, and BC; and the
  interaction among ABC

Follow through the steps below.

- *Have you read the* Limitations and Exceptions *section?*

- *Are all of these assumptions met for your study? If not, do not continue.*

- *Verbal equivalents for all notations are presented in the* Computations *section below. (See also* Note on Notation, *p. 133.)*

- *If you are experienced with computation for the analysis of variance, go directly to Summary Table 7.19 and proceed in your usual way. If you require step-by-step guidance, begin with Section B,* Recording the Data.

- *For convenience in simplifying notation, we have used designs with a fixed set of levels. If your design has more (or fewer) levels, simply expand (or contract) the design by continuing any particular step until all levels or level combinations have been accounted for.*

A. See Table 7.19.

---

[8] This analysis is also called "randomized block factorial design."

Table 7.19. Summary Table: Three-Way Within-Subjects ANOVA*

Source	df	SS = Sum of Squares	MS = Mean Square	F
A	$(a-1)$	$SS_A = \sum_{A=1}^{a} T_A^2/n_A - G^2/N$	$SS_A/(a-1)$	MS_A/MS_{AS}
B	$(b-1)$	$SS_B = \sum_{B=1}^{b} T_B^2/n_B - G^2/N$	$SS_B/(b-1)$	MS_B/MS_{BS}
C	$(c-1)$	$SS_C = \sum_{C=1}^{c} T_C^2/n_C - G^2/N$	$SS_C/(c-1)$	MS_C/MS_{CS}
S	$(s-1)$	$SS_S = \sum_{S=1}^{s} T_S^2/n_S - G^2/N$	(not computed)	
AB	$(a-1) \cdot$ $(b-1)$	$SS_{AB} = \sum_{AB=1}^{ab} T_{AB}^2/n_{AB} - SS_A - SS_B - G^2/N$	$SS_{AB}/(a-1)(b-1)$	MS_{AB}/MS_{ABS}
AC	$(a-1) \cdot$ $(c-1)$	$SS_{AC} = \sum_{AC=1}^{ac} T_{AC}^2/n_{AC} - SS_A - SS_C - G^2/N$	$SS_{AC}/(a-1)(c-1)$	MS_{AC}/MS_{ACS}
BC	$(b-1) \cdot$ $(c-1)$	$SS_{BC} = \sum_{BC=1}^{bc} T_{BC}^2/n_{BC} - SS_B - SS_C - G^2/N$	$SS_{BC}/(b-1)(c-1)$	MS_{BC}/MS_{BCS}
AS	$(a-1) \cdot$ $(s-1)$	$SS_{AS} = \sum_{AS=1}^{as} T_{AS}^2/n_{AS} - SS_A - SS_S - G^2/N$	$SS_{AS}/(a-1)(s-1)$	
BS	$(b-1) \cdot$ $(s-1)$	$SS_{BS} = \sum_{BS=1}^{bs} T_{BS}^2/n_{BS} - SS_B - SS_S - G^2/N$	$SS_{BS}/(b-1)(s-1)$	
CS	$(c-1) \cdot$ $(s-1)$	$SS_{CS} = \sum_{CS=1}^{cs} T_{CS}^2/n_{CS} - SS_C - SS_S - G^2/N$	$SS_{CS}/(c-1)(s-1)$	
ABC	$(a-1) \cdot$ $(b-1) \cdot$ $(c-1)$	$SS_{ABC} = \sum_{ABC=1}^{abc} T_{ABC}^2/n_{ABC} - SS_A - SS_B$ $- SS_C - SS_{AB} - SS_{AC} - SS_{BC} - G^2/N$	$SS_{ABC}/(a-1)(b-1)(c-1)$	MS_{ABC}/MS_{ABCS}

Table 7.19 (continued)

Source	df	SS = Sum of Squares	MS = Mean Square	F
ABS	$(a-1) \cdot$ $(b-1) \cdot$ $(s-1)$	$SS_{ABS} = \sum\limits_{ABS=1}^{abs} T^2_{ABS}/n_{ABS} - SS_A - SS_B$ $\quad - SS_S - SS_{AB} - SS_{AS} - SS_{BS}$ $\quad - G^2/N$	$SS_{ABS}/(a-1)(b-1)(s-1)$	
ACS	$(a-1) \cdot$ $(c-1) \cdot$ $(s-1)$	$SS_{ACS} = \sum\limits_{ACS=1}^{acs} T^2_{ACS}/n_{ACS} - SS_A - SS_C$ $\quad - SS_S - SS_{AC} - SS_{AS} - SS_{CS}$ $\quad - G^2/N$	$SS_{ACS}/(a-1)(c-1)(s-1)$	
BCS	$(b-1) \cdot$ $(c-1) \cdot$ $(s-1)$	$SS_{BCS} = \sum\limits_{BCS=1}^{bcs} T^2_{BCS}/n_{BCS} - SS_B - SS_C$ $\quad - SS_S - SS_{BC} - SS_{BS} - SS_{CS}$ $\quad - G^2/N$	$SS_{BCS}/(b-1)(c-1)(s-1)$	
ABCS	$(a-1) \cdot$ $(b-1) \cdot$ $(c-1) \cdot$ $(s-1)$	$SS_{ABCS} = SS_{\text{Total}} - SS_A - SS_B - SS_C$ $\quad - SS_S - SS_{AB} - SS_{AC} - SS_{BC}$ $\quad - SS_{AS} - SS_{BS} - SS_{CS} - SS_{ABC}$ $\quad - SS_{ABS} - SS_{ACS} - SS_{BCS}$	$SS_{ABCS}/(a-1)(b-1)(c-1)(s-1)$	
Total	$(N-1)$	$SS_{\text{Total}} = \sum X^2 - G^2/N$		

*Some statisticians recommend an alternative way of performing this analysis. Since the interaction error terms (*AS, BS, CS, ABS, ACS, BCS,* and *ABCS*) have few degrees of freedom and because there is no reason to expect them to provide an estimate of anything except random error, some statisticians recommend that these terms be pooled. This change provides a single-error term for all *F* ratios rather than seven different error terms. Pooling requires two changes in the computations as indicated. First, none of the interaction error terms (the interactions involving *S* listed above) are computed. (In every case, eliminate these terms completely from "Source" and all other columns of the Summary Table.) Instead, a pooled-error term is computed, using the following formula: $SS_{\text{Pooled Error}} = SS_{\text{Total}} - SS_A - SS_B - SS_C - SS_S - SS_{AB} - SS_{AC} - SS_{BC} - SS_{ABC}$; $df_{\text{Error}} = (s - 1)(abc - 1)$. Second, in the computation of the *F* ratios, instead of using the *MS* indicated in the denominator of the respective formulas, the new pooled-error term is used in every case. All of the other computations are identical.

B. RECORDING THE DATA

1. Record the data in an $a \times b \times c \times s$ raw-data table. For computational convenience, each individual row of numbers comprises data for only one subject. The major headings for the columns of this table, the A-factor headings, should each be divided into B-factor subheadings. Each of these is in turn divided into C-factor subheadings, as shown in Table 7.20. In the column margins are the totals for each of the ABC combinations. As a convenient case, we'll select for general coverage a three-factor design with two levels of each factor ($2 \times 2 \times 2 \times s$ table). Remember that each subject receives one score for each level of each factor (in this case, eight scores).

C. INITIAL STEPS

(These steps are spelled out in greater detail for the somewhat comparable analysis in the *Initial Steps* to the two-way within-subjects design.)

1. Construct subtotal tables to contain totals for each of the following sources: AS, BS, CS, ABS, ACS, BCS. Always make S the row factor: A, B, and C are column heads.
2. Begin a Summary Table. Include the column heads for this table exactly as they are indicated in Summary Table 7.19. Include the row heads (the items listed under "Source"), being sure that you use the appropriate *name of the factors* as well as the labels A, B, and so on. Complete the table as you proceed through the computations below. Each critical step is marked by the notation: "Record in Summary Table."
3. Compute the degrees of freedom (df), using the formulas in the df column of Table 7.19, and record them in your Summary Table.
4. There are 36 steps in this analysis. Since later computations refer back to steps by number, you will save time if you list the products from each step by number as you proceed. In addition, you can locate and correct errors in your computations more easily if you have recorded the intermediate steps systematically.

D. COMPUTATIONS

SUPERCHECK: No negative sums of squares (SS) may be obtained in the ANOVA. If you obtain a negative SS at any point, go back and check the preceding steps. Never go on if you have obtained a negative SS.

Table 7.20 Raw-Data Table: Three-Way Within-Subjects ($2 \times 2 \times 2 \times s$)

| | A_1 | | | | A_2 | | | | |
| | B_1 | | B_2 | | B_1 | | B_2 | | |
	C_1	C_2	C_1	C_2	C_1	C_2	C_1	C_2	
S_1	X_1	X_1	X_1	X_1	X_1	X_1	X_1	X_1	T_{S_1} (n_S)
S_2	X_2	X_2	X_2	X_2	X_2	X_2	X_2	X_2	T_{S_2} (n_S)
S_3	X_3	X_3	X_3	X_3	X_3	X_3	X_3	X_3	T_{S_3} (n_S)
S_4	X_4	X_4	X_4	X_4	X_4	X_4	X_4	X_4	T_{S_4} (n_S)
S_5	X_5	X_5	X_5	X_5	X_5	X_5	X_5	X_5	T_{S_5} (n_S)
S_6	X_6	X_6	X_6	X_6	X_6	X_6	X_6	X_6	T_{S_6} (n_S)
$\cdots$	$\cdots$	$\cdots$	$\cdots$	$\cdots$	$\cdots$	$\cdots$	$\cdots$	$\cdots$	$\cdots$
S_S	X_S	X_S	X_S	X_S	X_S	X_S	X_S	X_S	T_S (n_S)
	$T_{A_1 B_1 C_1}$ (n_{ABC})	$T_{A_1 B_1 C_2}$ (n_{ABC})	$T_{A_1 B_2 C_1}$ (n_{ABC})	$T_{A_1 B_2 C_2}$ (n_{ABC})	$T_{A_2 B_1 C_1}$ (n_{ABC})	$T_{A_2 B_1 C_2}$ (n_{ABC})	$T_{A_2 B_2 C_1}$ (n_{ABC})	$T_{A_2 B_2 C_2}$ (n_{ABC})	G (N)

*To simplify the notation in this table, each raw score is indicated only by the number of the subject who produced it. To be precise, each of these scores should have subscripts indicating ABC classification. The one in the upper left-hand corner, for example, should read $X_{A_1 B_1 C_1 S_1}$, and the one in the lower right-hand corner $X_{A_2 B_2 C_2 S_s}$.

SUPERCHECK: *Each step* from 1 to 15 requires that *every score* be counted (in one category or another) *every time.* Be sure that you systematically count all scores. A check on each of these steps can be made by summing all the totals for any step. *The resulting value must always be G* (see *Check on Computations*, p. 134).

1. a. Find $T_{A_1B_1C_1}$, $T_{A_1B_1C_2}$, $T_{A_1B_2C_1}$, $T_{A_1B_2C_2}$, $T_{A_2B_1C_1}$, $T_{A_2B_1C_2}$, $T_{A_2B_2C_1}$, $T_{A_2B_2C_2}$.
   Sum the scores in each ABC column of the $A \times B \times C \times S$ raw-data table. For convenience, record these values in the column margins.

   b. Find n_{ABC}. $[n_{ABC} = s]$
   n_{ABC}, the number of scores on which each of the ABC totals is based, equals s. For convenience, record n_{ABC}, in parentheses, in the column margin immediately below the appropriate total, in the $A \times B \times C \times S$ raw-data table.

2. a. Find $T_{A_1B_1S_1}$, $T_{A_1B_1S_2}$, and so on, $T_{A_1B_2S_1}$, $T_{A_1B_2S_2}$, and so on, $T_{A_2B_1S_1}$, $T_{A_2B_1S_2}$, and so on, $T_{A_2B_2S_1}$, $T_{A_2B_2S_2}$, and so on.
   For each subject, sum together all scores appearing in any one of the four AB combinations (by summing across levels of C). Record these totals in the cells of your $A \times B \times S$ subtotal table.

   b. Find n_{ABS}. $[n_{ABS} = c]$
   n_{ABS}, the number of scores on which each of the ABS totals is based, equals c. For convenience, record n_{ABS}, in parentheses, in the $A \times B \times S$ table immediately below the appropriate totals.

3. a. Find $T_{A_1C_1S_1}$, $T_{A_1C_1S_2}$, and so on, $T_{A_1C_2S_1}$, $T_{A_1C_2S_2}$, and so on, $T_{A_2C_1S_1}$, $T_{A_2C_1S_2}$, and so on, $T_{A_2C_2S_1}$, $T_{A_2C_2S_2}$, and so on.
   For each subject, sum together all scores in any one of the four AC combinations (by summing across the levels of B). Record these in the cells of your $A \times C \times S$ table.

   b. Find n_{ACS}. $[n_{ACS} = b]$
   n_{ACS}, the number of scores on which each of the ACS totals is based, equals b. For convenience, record n_{ACS}, in parentheses, in the $A \times C \times S$ table immediately below the appropriate total.

4. a. Find $T_{B_1C_1S_1}$, $T_{B_1C_1S_2}$, and so on, $T_{B_1C_2S_1}$, $T_{B_1C_2S_2}$, and so on, $T_{B_2C_1S_1}$, $T_{B_2C_1S_2}$, and so on, $T_{B_2C_2S_1}$, $T_{B_2C_2S_2}$, and so on.

For each subject, sum together all scores in any one of the four BC combinations by summing across the levels of A. Record these totals in the cells of your $B \times C \times S$ table.

b. Find n_{BCS}. [$n_{BCS} = a$]

n_{BCS}, the number of scores on which each of the BCS totals is based, equals a. For convenience, record n_{BCS}, in parentheses, in the $B \times C \times S$ table immediately below the appropriate total.

5. a. From the $A \times B \times S$ table, find $T_{A_1B_1}, T_{A_1B_2}, T_{A_2B_1}, T_{A_2B_2}$. Sum together all scores in each of the four AB combinations; that is, sum *across* all subjects. Record these totals in the appropriate column margins of your $A \times B \times S$ table.

b. Find n_{AB}. [$n_{AB} = c \times s$]

n_{AB}, the number of scores on which each of the AB totals is based, equals $c \times s$. For convenience, record n_{AB}, in parentheses, in the column margins of the $A \times B \times S$ table immediately below the appropriate total.

6. a. From the $A \times C \times S$ table, find $T_{A_1C_1}, T_{A_1C_2}, T_{A_2C_1}, T_{A_2C_2}$. Sum together all scores in each of the four AC combinations; that is, sum *across* all subjects. Record these totals in the appropriate column margins of your $A \times C \times S$ table.

b. Find n_{AC}. [$n_{AC} = b \times s$]

n_{AC}, the number of scores on which each of the AC totals is based, equals $b \times s$. For convenience, record n_{AC}, in parentheses, in the column margins of the $A \times C \times S$ table immediately below the appropriate total.

7. a. From the $B \times C \times S$ table, find $T_{B_1C_1}, T_{B_1C_2}, T_{B_2C_1}, T_{B_2C_2}$. Sum together all scores in each of the four BC combinations; that is, sum *across* all subjects. Record these totals in the appropriate column margins of your $B \times C \times S$ table.

b. Find n_{BC}. [$n_{BC} = a \times s$]

n_{BC}, the number of scores on which each of the BC totals is based, equals $a \times s$. For convenience, record n_{BC}, in parentheses, in the column margins of the $B \times C \times S$ table immediately below the appropriate total.

8. a. From the $A \times B \times S$ table, find $T_{A_1S_1}, T_{A_1S_2}$, and so on, $T_{A_2S_1}, T_{A_2S_2}$, and so on.

For each subject, sum together all scores appearing in each of the two levels of factor A (by summing *across* the levels of B). Record these totals in the cells of your $A \times S$ table.

b. Find n_{AS}. [$n_{AS} = b \times c$]

Determine the number of scores on which each of the AS

totals is based. For each total, record n_{AS}, in parentheses, in
the $A \times S$ table immediately below the appropriate total.

9. a. From the $B \times C \times S$ table, find $T_{B_1 S_1}$, $T_{B_1 S_2}$, and so on,
$T_{B_2 S_1}$, $T_{B_2 S_2}$, and so on.
For each subject, sum together all scores in each of the two
levels of factor B (by summing *across* the levels of C). Record
these totals in the cells of your $B \times S$ table.

b. Find n_{BS}. [$n_{BS} = a \times c$]
n_{BS}, the number of scores on which each of the BS totals is
based, equals $a \times c$. For convenience, record n_{BS}, in parentheses, in the $B \times S$ table immediately below the appropriate
total.

10. a. From the $B \times C \times S$ table, find $T_{C_1 S_1}$, $T_{C_1 S_2}$, and so on,
$T_{C_2 S_1}$, $T_{C_2 S_2}$, and so on.
For each subject, sum together all scores in each of the two
levels of factor C (by summing *across* the levels of B). Record
these totals in the cells of your $C \times S$ table.

b. Find n_{CS}. [$n_{CS} = a \times b$]
n_{CS}, the number of scores on which each of the CS totals is
based, equals $a \times b$. For convenience, record n_{CS}, in parentheses, in the $C \times S$ table immediately below the appropriate
total.

11. a. From the $A \times S$ table, find T_{A_1}, T_{A_2}.
Sum the scores in each column of your $A \times S$ table; that is
sum *across* all subjects. Record these totals in the column
margins.

b. Find n_A. [$n_A = b \times c \times s$]
n_A, the number of scores on which each of the A totals is
based, equals $b \times c \times s$. For convenience, record n_A, in
parentheses, immediately below the appropriate total in the
$A \times S$ table.

12. a. From the $B \times S$ table, find T_{B_1}, T_{B_2}.
Sum the scores in each column of your $B \times S$ table; that is,
sum *across* all subjects. Record these totals in the column
margins.

b. Find n_B. [$n_B = a \times c \times s$]
n_B, the number of scores on which each of the B totals is
based, equals $a \times c \times s$. For convenience, record n_B, in
parentheses, immediately below the appropriate total in the
$B \times S$ table.

13. a. From the $C \times S$ table, find T_{C_1}, T_{C_2}.

Sum the scores in each column of your $C \times S$ table; that is, sum *across* all subjects. Record these totals in the column margins.

b. Find n_C. [$n_C = a \times b \times s$]

n_C, the number of scores on which each of the C totals is based, equals $a \times b \times s$. For convenience, record n_C, in parentheses, immediately below the appropriate total in the $C \times S$ table.

14. a. From the $A \times B \times C \times S$ table, find $T_{S_1}, T_{S_2}, T_{S_3}, \ldots T_{S_n}$.

Sum together all scores in each row of your $A \times B \times C \times S$ raw-data table; that is, for each subject, sum across all levels of factors A, B, and C. Record these totals in the row margins of your $A \times B \times C \times S$ raw-data table.

b. Find n_S. [$n_S = a \times b \times c$]

n_S, the number of scores on which each of the S totals is based, equals $a \times b \times c$. For convenience, record n_S, in parentheses, immediately below the appropriate total in the $A \times B \times C \times S$ table.

15. a. Find G.

Sum all the scores in your $A \times B \times C \times S$ raw-data table. Record G at the bottom of the $A \times B \times C \times S$ table.

CHECK: $T_{A_1} + T_{A_2} = T_{B_1} + T_{B_2} = T_{C_1} + T_{C_2} = G$

$T_{S_1} + T_{S_2} + T_{S_3} + T_{S_4} + \cdots + T_{S_n} = G$

b. Find N.

Count the total number of scores in the $A \times B \times C \times S$ raw-data table. Record N in parentheses next to G.

CHECK: $N = a \times b \times c \times s$

16. Find G^2/N.

Square G and divide the obtained value by N.

17. Find ΣX^2.

Square each score in the $A \times B \times C \times S$ raw-data table; sum the squared values.

18. Find $SS_{\text{Total}} = \Sigma X^2 - G^2/N$.

Subtract: #17 − #16. Record in Summary Table.

19. Find $SS_S = \Sigma T_S^2/n_S - G^2/N$.

a. Square each of the values from #14a, sum the squared values, and divide by the number of scores for each subject, n_S (#14b).

b. From this value subtract #16. Record in Summary Table.

20. Find $SS_A = \Sigma T_A^2/n_A - G^2/N$.
    a. Square each of the values from #11a, sum the squared values, and divide by n_A (#11b).
    b. From this value subtract #16. Record in Summary Table.

21. Find $SS_B = \Sigma T_B^2/n_B - G^2/N$.
    a. Square each of the values from #12a, sum the squared values, and divide by n_B (#12b).
    b. From this value subtract #16. Record in Summary Table.

22. Find $SS_C = \Sigma T_C^2/n_C - G^2/N$.
    a. Square each of the values from #13a, sum the squared values, and divide by n_C (#13b).
    b. From this value subtract #16. Record in Summary Table.

23. Find $SS_{AB} = \Sigma T_{AB}^2/n_{AB} - SS_A - SS_B - G^2/N$.
    a. Square each of the values from #5a, sum the squared values, and divide by n_{AB} (#5b).
    b. From this value, subtract #20b, #21b, and #16. Record in Summary Table.

24. Find $SS_{AC} = \Sigma T_{AC}^2/n_{AC} - SS_A - SS_C - G^2/N$.
    a. Square each of the values from #6a, sum the squared values, and divide by n_{AC} (#6b).
    b. From this value subtract #20b, #22b, and #16. Record in Summary Table.

25. Find $SS_{BC} = \Sigma T_{BC}^2/n_{BC} - SS_B - SS_C - G^2/N$.
    a. Square each of the values from #7a, sum the squared values, and divide by n_{BC} (#7b).
    b. From this value subtract #21b, #22b, and #16. Record in Summary Table.

26. Find $SS_{AS} = \Sigma T_{AS}^2/n_{AS} - SS_A - SS_S - G^2/N$.
    a. Square each of the values from #8a, sum the squared values, and divide by n_{AS} (#8b).
    b. From this value, subtract #20b, #19b, and #16. Record in Summary Table.

27. Find $SS_{BS} = \Sigma T_{BS}^2/n_{BS} - SS_B - SS_S - G^2/N$.
    a. Square each of the values from #9a, sum the squared values, and divide by n_{BS} (#9b).
    b. From this value subtract #21b, #19b, and #16. Record in Summary Table.

28. Find $SS_{CS} = \Sigma T_{CS}^2/n_{CS} - SS_C - SS_S - G^2/N$.
    a. Square each of the values from #10a, sum the squared values, and divide by n_{CS} (#10b).
    b. From this value subtract #22b, #19b, and #16. Record in Summary Table.

29. Find $SS_{ABC} = \Sigma T_{ABC}^2/n_{ABC} - SS_A - SS_B - SS_C - SS_{AB} - SS_{AC} - SS_{BC} - G^2/N$.
    a. Square each of the values from #1a, sum the squared values, and divide by n_{ABC} (#1b).
    b. From this value subtract #20b, #21b, #22b, #23b, #24b, #25b, and #16. Record in Summary Table.
30. Find $SS_{ABS} = \Sigma T_{ABS}^2/n_{ABS} - SS_A - SS_B - SS_S - SS_{AB} - SS_{AS} - SS_{BS} - G^2/N$.
    a. Square each of the values from #2a, sum the squared values, and divide by n_{ABS} (#2b).
    b. From this value subtract #20b, #21b, #19b, #23b, #26b, #27b, and #16. Record in Summary Table.
31. Find $SS_{ACS} = \Sigma T_{ACS}^2/n_{ACS} - SS_A - SS_C - SS_S - SS_{AC} - SS_{AS} - SS_{CS} - G^2/N$.
    a. Square each of the values from #3a, sum the squared values, and divide by n_{ACS} (#3b).
    b. From this value subtract #20b, #22b, #19b, #24b, #26b, #28b, and #16. Record in Summary Table.
32. Find $SS_{BCS} = \Sigma T_{BCS}^2/n_{BCS} - SS_B - SS_C - SS_S - SS_{BC} - SS_{BS} - SS_{CS} - G^2/N$.
    a. Square each of the values from #4a, sum the squared values, and divide by n_{BCS} (#4a).
    b. From this value subtract #21b, #22b, #19b, #25b, #27b, #28b, and #16. Record in Summary Table.
33. Find SS_{ABCS}.
    Subtract: #18 − #19b − #20b − #21b − #22b − #23b − #24b − #25b − #26b − #27b − #28b − #29b − #30b − #31b − #32b. Record in Summary Table.
34. Compute the 15 mean squares indicated in Summary Table 7.19 by dividing each sum of squares by its appropriate (corresponding) *df*. Record in Summary Table.
35. Find the seven *F*s by dividing mean squares by the appropriate error term (mean-square error), as indicated in Summary Table 7.19. Record in Summary Table.
36. Evaluate computed *F*s by consulting the *F* table, Appendix 3, using the appropriate *df*.

D. FURTHER PROCEDURES

1. Interpretation: see Chapter 9, *Introduction* and Section C.
2. Specific-comparison tests: see Chapter 10, *Introduction* and Section C.
3. Strength-of-association measure: see Chapter 11, *Introduction* and Section C.

EXTENSIONS OF

WITHIN-SUBJECTS

ANOVA

● *If you are experienced with computations for the analysis of variance, the following recommendations will probably provide a sufficient guideline. If you require step-by-step guidance, return to the three-way-within-subjects ANOVA, p. 192, for more specific instructions.*

● *Reread the section,* The Common Pattern for All Analysis of Variance, *p. 136, before beginning to set up your extension.*

A. DEVELOPING THE SUMMARY TABLE

1. Set up a Summary Table with the usual headings: *Source, df, SS, MS,* and *F.*
2. a. In listing your sources in a within-subjects table, the general principle is to systematically list your variables (main effects) both by name and by label (*A, B, C,* and so on). Remember that, in listing your sources, subjects (*S*) must be included.
   b. List your two-way interactions, three-way interactions, and so on. The two-way interactions comprise all possible two-way combinations of the variables *A, B, C,* and so on. The three-way interactions comprise all possible three-way combinations of the variables. There are, however, two correct ways of performing the analysis. One of these (see note to Table 7.19) permits you to pool all interactions involving subjects (*S*) to form a single error term that is used in forming all *F* ratios. If you use this method, you list among your sources *no* interactions involving *S* as a variable. The second method (unpooled) uses the interaction of the critical variables with *S* as the error term in forming the *F* ratio; for example, $F_{AB} = MS_{AB}/MS_{ABS}$. This method involves listing among your sources all interactions, including those involving *S*. As a check, we list in Table 7.21 the number of main effects and interactions for analyses with up to seven variables. Enter the table including *S* as a variable if you plan to use an unpooled-error term. Do *not* include *S* as a variable if you plan to use a pooled-error term. (A glance at Table 7.21 may prove sobering. Any of these extensions means a lot of work, but the pooled-error-term method clearly involves many fewer steps.)

Table 7.21

Number of Variables*	Total Number of Interactions	2-way	3-way	4-way	5-way	6-way	7-way
1	(0)						
2	(1)	1					
3	(4)	3	1				
4	(11)	6	4	1			
5	(26)	10	10	5	1		
6	(57)	15	20	15	6	1	
7	(120)	21	35	35	21	7	1

*Include subjects as a variable if you plan to use the unpooled method. Do *not* include subjects as a variable if you plan to use the pooled method

3. In addition, you must include the source: Total. There is usually no error term in these designs except in the unusual case in which you have two scores on every subject in every condition.
4. Degrees of freedom are always computed in exactly the same way. For the main effects, df are always one less than the number of levels for the particular factor. For the interactions, df are always the cross-products of the df for the appropriate main effects. For example, df_{BC} would be $(b - 1)(c - 1)$, while df_{BDS} would be $(b - 1)(d - 1)(s - 1)$. df_{Total} is always $N - 1$. Note that the sum of all the dfs in any table equals df_{Total} or $N - 1$.

B. COMPUTATIONS

For specific computational procedures, read *The Common Pattern for All Analysis of Variance*, p. 136; then, if you require further guidance, follow through the steps (including additional steps at the appropriate places) of the two-way-within-subjects or the three-way-within-subjects ANOVA.

Description:

One independent variable
Two levels only
Between-subjects
| t TEST | *n*s need *not* be equal
Strength-of-association measure available
Specific comparisons not needed
Other appropriate analysis: one-way between-
  subjects ANOVA

Follow through the steps below.

● *Have you read the* Limitations and Exceptions *section?*

● *Are all of these assumptions met for your study? If not, do not
continue.*

● *Verbal equivalents for all notations are presented in the* Computations *section below.*

● *If you are experienced with computations for the* t *test, go directly
to the computing formula below. If you require step-by-step guidance,
begin with Section B,* The Example.

A. COMPUTING FORMULA

$$t = \frac{\bar{X}_1 - \bar{X}_2}{\sqrt{\left(\frac{\left(\Sigma X_1^2 - \frac{(\Sigma X_1)^2}{n_1}\right) + \left(\Sigma X_2^2 - \frac{(\Sigma X_2)^2}{n_2}\right)}{(n_1 + n_2 - 2)}\right)\left(\frac{n_1 + n_2}{(n_1)(n_2)}\right)}}$$

where X_1 = any score from Group 1,
       $\bar{X}_1$ = the mean of Group 1,
       n_1 = the number of subjects in Group 1,
       X_2 = any score from Group 2,
       $\bar{X}_2$ = the mean of Group 2,
       n_2 = the number of subjects in Group 2,
       N = total number of subjects.

t is evaluated with $N - 2\ df$. See Appendix 4.

B. THE EXAMPLE

1. A college professor who used programmed instruction to teach
   his introductory psychology class wished to determine whether
   written or oral·testing was more motivating for his students. He

randomly assigned ten of his students to receive oral testing. The
remaining ten students were assigned to a written-testing condition. At the end of three weeks, he counted the number of
units that each student had successfully completed.
2. *Null Hypothesis*
   a. *General.* The sample means were drawn from populations
      having the same means.
   b. *Specific.* Mean test performance for both oral and written test
      conditions were drawn from populations having the same
      means.
3. *Steps through the Branching Program for This Example*
   a. *Statistical technique needed:* **test of significance**
   b. *Type of data:* **score data.** Each subject has a score consisting
      of the number of units successfully completed during the
      three-week period.
   c. *Between, within, mixed:* **between.** Different subjects served in
      the written and oral conditions and were measured only once.
   d. *Number of independent variables:* **one.** Type of testing
   e. *Number of levels of the independent variable:* **two.** Written
      and oral testing

C. RECORDING THE DATA

Record the data in a table with two columns, one for each level of
the independent variable. Label the two columns. At the bottom of
the table, record ΣX, ΣX^2, and n for each group (see Table 7.22).

Table 7.22

Oral Testing	Written Testing
2	3
2	1
3	2
4	1
3	2
2	1
5	0
3	3
4	1
5	2
$\Sigma X_1 = 33$	$\Sigma X_2 = 16$
$\Sigma X_1^2 = 121$	$\Sigma X_2^2 = 34$
$n_1 = 10$	$n_2 = 10$

D. COMPUTATIONS

1. Obtain n_1.
   Count the number of subjects in Group 1.

$$n_1 = 10$$

2. Obtain n_2.
   Count the number of subjects in Group 2.

$$n_2 = 10$$

3. Obtain ΣX_1.
   Sum all of the scores for Group 1.

$$2 + 2 + \cdots 5 = 33$$

4. Obtain ΣX_2.
   Sum all of the scores for Group 2.

$$3 + 1 + \cdots 2 = 16$$

5. Obtain ΣX_1^2.
   Square all of the scores for Group 1, then add them.

$$4 + 4 + \cdots + 25 = 121$$

6. Obtain ΣX_2^2.
   Square all of the scores for Group 2, then add them.

$$9 + 1 + \cdots + 4 = 34$$

7. Obtain $\Sigma X_1^2 - (\Sigma X_1)^2/n_1$.
   Square #3; divide by #1. Subtract this figure from #5.

$$121 - (33)^2/10 = 121 - 108.9 = 12.1$$

8. Obtain $\Sigma X_2^2 - (\Sigma X_2)^2/n_2$.
   Square #4; divide by #2. Subtract this figure from #6.

$$34 - (16)^2/10 = 34 - 25.6 = 8.4$$

9. Obtain $\dfrac{(\Sigma X_1^2 - (\Sigma X_1)^2/n_1) + (\Sigma X_2^2 - (\Sigma X_2)^2/n_2)}{(n_1 + n_2 - 2)}$.
   Add #7 and #8. Divide this figure by $n_1 + n_2 - 2$ (#1 + #2 − 2).

$$\frac{12.1 + 8.4}{18} = \frac{20.5}{18} = 1.139$$

10. Obtain $\dfrac{n_1 + n_2}{(n_1)(n_2)}$.
    Add #1 and #2. Divide this sum by the product of #1 and #2.

$$\frac{10 + 10}{(10)(10)} = \frac{20}{100} = .2$$

11. Obtain $\sqrt{\left(\dfrac{\left(\Sigma X_1^2 - \dfrac{(\Sigma X_1)^2}{n_1}\right) + \Sigma X_2^2 - \dfrac{(\Sigma X_2)^2}{n_2}}{(n_1 + n_2 - 2)}\right)\left(\dfrac{n_1 + n_2}{(n_1)(n_2)}\right)}$.
    Multiply #9 by #10. Obtain the square root.

$$\sqrt{(1.139)(.2)} = \sqrt{0.228} = .477$$

12. Obtain $\bar{X}_1 = \dfrac{\Sigma X_1}{n_1}$.
    Divide #3 by #1.

$$\frac{33}{10} = 3.3$$

13. Obtain $\bar{X}_2 = \dfrac{\Sigma X_2}{n_2}$.
    Divide #4 by #2.

$$\frac{16}{10} = 1.6$$

14. Obtain $\bar{X}_1 - \bar{X}_2$.
    Subtract #13 from #12.

$$3.3 - 1.6 = 1.7$$

15. Obtain t.
    Divide #14 by #11.

$$\frac{1.7}{.477} = 3.564$$

16. Obtain the critical value of t for $\alpha = .05$ and $df = N - 2$ from
    Appendix 4. If t is negative, drop the sign.

> The critical value of t for $\alpha = .05$ for
> $N - 2 = 20 - 2 = 18$ df is 2.101.

17. a. If $t_{\text{Computed}} \geq t_{\text{Tabled}}$, the difference between the two groups
       is significant. Reject the null hypothesis.
    b. If $t_{\text{Computed}} < t_{\text{Tabled}}$, the difference between the groups is not
       significant. Do not reject the null hypothesis.

> $3.564 > 2.101$. Therefore the difference between the groups
> is significant. Reject the null hypothesis.

E. INTERPRETATION OF THE EXAMPLE

1. *Statistical conclusion.* If these observations were drawn randomly
   from populations having the same means, means as discrepant as
   these would occur less than 5% of the time by chance. Therefore
   the null hypothesis should be rejected.
2. *Researcher's conclusion.* Subjects receiving oral testing complete
   more units than subjects who receive written testing.

NOTE: The reader is invited to apply a one-way between-subjects ANOVA to these data to satisfy himself that (1) analysis of variance is computationally simpler and that (2) t^2 does indeed equal F.

F. FURTHER PROCEDURES

1. Interpretations: see Chapter 9, *Introduction* and Section C.
2. Specific comparisons: not needed.
3. Strength-of-association measure: see Chapter 11, *Introduction* and Section C.

Description:

One independent variable
Two levels only
ns must be equal (see p. 130)
Within-subjects
Strength-of-association
    measure available
Specific comparisons not
    needed
Other appropriate analysis:
    one-way within-subjects
    ANOVA

---

t **TEST FOR**

CORRELATED MEANS

Follow through the steps below.

- *Have you read the* Limitations and Exceptions *section?*

- *Are all of these assumptions met for your study? If not, do not continue.*

- *Verbal equivalents for all notations are presented in the* Computations *section below.*

- *If you are experienced with computations for the* t *test for correlated means, go directly to the computing formula below. If you require step-by-step guidance, begin with Section B,* The Example.

A. COMPUTING FORMULA

$$t = \frac{\bar{X}_D}{s_D / \sqrt{N}}$$

where $\bar{X}_D$ = the mean of the differences between Measurement 1
            and Measurement 2,
    s_D = the standard deviation of the difference scores, and
    N = the total number of subjects.
t is evaluated with $N - 1$ df. See Appendix 4.

B. THE EXAMPLE

1. (We use the same example that we used to illustrate the *t* test for between-subjects designs, but we have converted it to a within-subjects design.) A college professor who used programmed instruction to teach his introductory psychology class wished to determine whether written or oral testing was more motivating for his students. Since he had only ten students in his class, he decided that each student would participate in both the oral and the written testing conditions. Half of the students received oral testing for the first three weeks and written testing for the next three weeks. The other half received written testing for the first three weeks and oral testing for the next three (a counter-balanced design). However, the professor was not interested in testing the effects of order but was concerned only with the difference between oral and written testing.

2. *Null Hypothesis*
   a. *General.* The sample means were drawn from populations having the same means.
   b. *Specific.* Mean test performance for both oral and written test conditions were drawn from populations having the same means.

3. *Steps through the Branching Program for This Example*
   a. *Statistical technique needed:* **test of significance**
   b. *Type of data:* **score data.** Each subject has two scores, consisting of the number of units successfully completed under oral testing and the number of units successfully completed under written testing.
   c. *Between, within, mixed:* **within.** The same subjects serve in both the written and the oral testing conditions.
   d. *Number of independent variables:* **one.** Type of testing
   e. *Number of levels of the independent variable:* **two.** Written and oral testing

C. RECORDING THE DATA

Record the data in a table with four columns. In the first column, give the name or identifying number of the subject. In the next two columns, record the scores for Measurements 1 and 2, respectively. In the final column (D), record the difference between each subject's two scores. At the bottom of the last column, record ΣD, ΣD^2, and N (see Table 7.23).

Table 7.23

Subject Number	Measurement 1 Oral Testing	Measurement 2 Written Testing	D (1 − 2)
1	2	3	−1
2	2	1	+1
3	3	2	+1
4	4	1	+3
5	3	2	+1
6	2	1	+1
7	5	0	+5
8	3	3	0
9	4	1	+3
10	5	2	+3

$$\Sigma D = +17$$
$$\Sigma D^2 = 57$$
$$N = 10$$

D. COMPUTATIONS

1. Obtain N.
   Count the number of subjects in the study.

$$\boxed{N = 10}$$

2. Obtain ΣD.
   Sum the difference scores. ΣD may be positive or negative.

$$\boxed{-1 + 1 + \cdots + 3 = +17}$$

3. Obtain ΣD^2.
   Square each difference score; then add them.

$$\boxed{1 + 1 + \cdots + 9 = 57}$$

4. Obtain $s_D = \sqrt{\dfrac{\Sigma D^2 - (\Sigma D)^2/N}{N - 1}}$

   Square #2; divide by #1. Subtract this figure from #3. Divide by $N - 1$. Extract the square root.

$$\boxed{\sqrt{\frac{57 - (17)^2/10}{9}} = \sqrt{\frac{57 - 28.9}{9}} = \sqrt{\frac{28.1}{9}} = \sqrt{3.1222} = 1.767}$$

5. Obtain $\bar{X}_D = \dfrac{\Sigma D}{N}$.

Divide #2 by #1. $\bar{X}_D$ may be positive or negative.

$$\boxed{\dfrac{17}{10} = 1.7}$$

6. Obtain $s_D/\sqrt{N}$.
   Divide #4 by $\sqrt{N}$.

$$\boxed{1.767/\sqrt{10} = 1.767/3.163 = .559}$$

7. Obtain t.
   Divide #5 by #6.

$$\boxed{1.7/.559 = 3.04}$$

8. Obtain the critical value of t for $\alpha = .05$ and $df = N - 1$ from Appendix 4. If t is negative, drop the sign.

$$\boxed{\text{The critical value of } t \text{ for } \alpha = .05 \text{ and } df = N - 1 = 8 \text{ is } 2.306.}$$

9. a. If $t_{\text{Computed}} \geq t_{\text{Tabled}}$, the difference between the two measures is significant. Reject the null hypothesis.
   b. If $t_{\text{Computed}} < t_{\text{Tabled}}$, the difference between the two measures is not significant. Do not reject the null hypothesis.

$$\boxed{\begin{array}{c} 3.04 > 2.306. \text{ Therefore the difference between the two} \\ \text{measures is significant. Reject the null hypothesis.} \end{array}}$$

E. INTERPRETATION OF THE EXAMPLE

1. *Statistical conclusion.* If these observations were drawn randomly from populations having the same means, means as discrepant as these would occur less than 5% of the time by chance. Therefore the null hypothesis should be rejected.

2. *Researcher's conclusion.* When subjects receive oral testing, they complete more units than when they receive written testing.

F. FURTHER PROCEDURES

1. Interpretations: see Chapter 9, *Introduction* and Section C.
2. Specific comparisons: not needed.
3. Strength-of-association measure: see Chapter 11, *Introduction* and Section C.

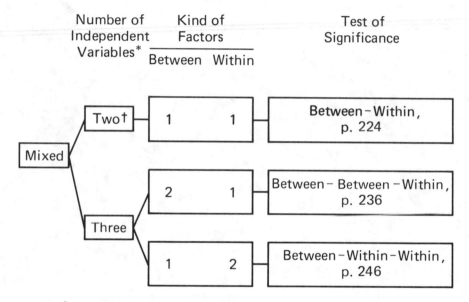

Number of Independent Variables*	Kind of Factors		Test of Significance
	Between	Within	
Two†	1	1	Between–Within, p. 224
Three	2	1	Between–Between–Within, p. 236
	1	2	Between–Within–Within, p. 246

Mixed

* All tests in this chapter will handle any number of levels of the independent variables (from two to the maximum number that the researcher might choose).

† At least two independent variables are required for a mixed design.

CHAPTER EIGHT

SCORE DATA: MIXED DESIGNS

LIMITATIONS AND EXCEPTIONS

The branching program on p. 120 makes three assumptions about the research design. These analyses are appropriate only if each of these is true:

1. *Fixed-effects model.* The design is a fixed-effects model for analysis of variance. This means that the levels of the independent variable(s) have been arbitrarily chosen by the researcher and that no generalizations are to be made beyond the levels that are studied. The distinction between fixed- and random-effects models is spelled out in Chapter 7, p. 128.

2. *Factorial combination.* All designs employing more than one independent variable have factorially combined the several independent variables. See Chapter 2, p. 9, for a definition of factorial combination.

3. *Equal number of scores.* There are an equal number of scores in each treatment or treatment combination. If you have an unequal number of scores, adopt one of the solutions suggested on p. 131.

Simple analysis-of-variance designs involve either administering all levels of the treatment or treatment combinations to different groups of subjects (see Table 8.1) or administering all levels of the treatment combinations to the same group of subjects (see Table 8.2).

Table 8.1. Between-Subjects (Independent) Design: 3 × 3 with Nine Groups of Subjects

Factor B (Between)

	B_1	B_2	B_3
A_1	Group 1	Group 2	Group 3
A_2	Group 4	Group 5	Group 6
A_3	Group 7	Group 8	Group 9

Factor A (Between)

Table 8.2. Within-Subjects (Repeated) Design: 3 × 3 with One Group of Subjects

Factor B (Within)

	B_1	B_2	B_3
A_1	Group 1	Group 1	Group 1
A_2	Group 1	Group 1	Group 1
A_3	Group 1	Group 1	Group 1

Factor A (Within)

For complex and mixed designs there are at least two factors, and a more complicated assignment of subjects to conditions prevails. There are a large number of complex designs;[1] we shall be able to examine only a few. Again, the limitations imposed by the length of this book reinforce our belief that researchers who design complex studies probably do not require a simplified manual to perform analyses.

We shall define three kinds of complex and mixed designs. *Only the first kind of these designs is covered in this manual.* However, you should read each of the descriptions to be sure you have properly classified your design. (We have deliberately provided brief, simple

[1] An excellent and relatively simple discussion of this topic appears in Kirk (1972), Chapter 7.2, Classification of ANOVA Designs, pp. 241–260.

descriptions of these design types. Any of these designs, however, may take on extremely complex—and sometimes almost unrecognizable—forms.)

1. *Between-subjects and within-subjects factors.* The simplest kind of mixed design is one that includes at least one between-subjects and one within-subjects factor. Any number (one or greater) of either kind of factor may be involved. Here, for simplicity, we illustrate a design with one between-subjects and one within-subjects factor. Note the assignment of groups of subjects in this 3 × 3 design. Each

Table 8.3. Mixed Design: 3 × 3 with Three Groups of Subjects

Factor B (Within)

		B_1	B_2	B_3
Factor A (Between)	A_1	Group 1	Group 1	Group 1
	A_2	Group 2	Group 2	Group 2
	A_3	Group 3	Group 3	Group 3

subject (or group of subjects) serves in $a = 3$ treatment conditions. Compare this design with the 3 × 3 between-subjects ANOVA and the 3 × 3 within-subjects ANOVA presented earlier. In the between-subjects ANOVA, every subject serves in *only one* treatment combination. In the within-subjects ANOVA, every subject serves in *all* treatment combinations. In the between-within ANOVA, every subject serves in *all levels* of the within factor (*B*), but in *only one level* of the between factor (*A*).

The simplest designs of this type are those with only one within-subjects factor and one or more between-subjects factors. The number of computational steps, *but not the complexity of the calculations*, increases with the number of between-subjects factors. As the number of within-subjects factors increases, the complexity of the analyses also increases.

2. *Hierarchically nested designs.* A nested design is a complex design in which levels of one factor are completely subsumed (nested) beneath another factor. (This, of course, means that the factors are not factorially combined.) A simple example may make this type of design clear. Suppose that you wish to study three kinds of psychotherapy, *A*, *B*, and *C*. This, of course, is a simple one-way between-

subjects design with a single factor, type of psychotherapy. But suppose that four hospitals are using treatment A, four are using treatment B, and four are using treatment C. There might be differences in hospitals, *per se*, that you wished to assess. If you wish to analyze the effects of hospitals, *per se*, you have a design that is neither between- nor within-subjects, since the factor of hospitals is not factorially combined with type of psychotherapy (see Table 8.4).

Table 8.4. Nested Design: 3 X 4 with Four Different Groups Nested Within Each of Three Factors

Psychotherapy A	*Psychotherapy B*	*Psychotherapy C*
Hospital 1	Hospital 5	Hospital 9
Hospital 2	Hospital 6	Hospital 10
Hospital 3	Hospital 7	Hospital 11
Hospital 4	Hospital 8	Hospital 12

3. *Latin square designs.* Latin squares are ordinarily used to control the order of conditions assigned to subjects in within-subjects designs. For example, you may have four treatment conditions and may wish to avoid confounding treatment conditions with order. Latin square designs make it possible to assign the order of the treatments (A_1, A_2, A_3, A_4) to subjects so that treatment condition and order are not confounded. Furthermore, the effects of both treatment condition and order may be independently assessed. Table 8.5 shows subjects as the row variables and order as the column variables. It may be seen that Subject 1 serves in treatments A_3, A_4, A_1, and A_2 in turn. Subject 2 serves in the order A_1, A_2, A_4, and A_3, and so on. Note that each treatment occurs only once in each row and column. This is a characteristic of all Latin square designs.[2]

Table 8.5. Latin Square Design: Subjects Are Assigned Four Treatment Conditions in One of Four Orders

	Order			
	1	2	3	4
S_1	A_3	A_4	A_1	A_2
S_2	A_1	A_2	A_4	A_3
S_3	A_4	A_3	A_2	A_1
S_4	A_2	A_1	A_3	A_4

(Subjects)

[2]We shall not discuss Latin square designs further (or present representative analyses), but a discussion can be found in Edwards (1968), p. 254.

NOTES ON THE ANALYSES AND WORKED EXAMPLES

Change in format of presentation of designs. If you have followed through the procedures in the earlier portions of this manual, we must warn you that the format of this chapter differs from that of the earlier chapters. The increase in length of the problems and their increased complexity has precluded our spelling out all of the steps for each analysis. We have provided a Summary Table and then briefly summarized a reasonable sequence of steps. If you have difficulty with the computations, refer back to earlier analyses. Appropriate analyses for reference have been suggested. We hope that we have effected a reasonable compromise between offering dozens of pages of step-by-step description and leaving you completely in the lurch when you begin to perform your computations. In addition, we have omitted examples for designs in this chapter except for the simplest, the between-within design.

Note that we have covered only the two-way and the three-way mixed-design analyses. All higher-order designs are logical extensions of these, and, as we have already indicated, the addition of between-subjects factors does not affect the complexity of the analyses. We recognize that it is more difficult to extend the principles involved when within-subject factors are added to the design. Such complex designs, however, are used relatively rarely.

Accompanying each analysis is a brief description that lists in a summary form when the analysis may be used, special requirements, and so forth.

Each analysis also includes a reference to the chapters on interpretation, specific comparisons, and strength of association.

Description:

Two independent variables

Any number of levels of
each variable

One within-subjects
(repeated) factor AND
one between-subjects
(independent) factor

ns must be equal (see
p. 130)

Fixed model only

Specific comparisons and
strength-of-association
tests available

Tests for differences
among levels of A,
among levels of B, and
interaction between A
and B

BETWEEN-WITHIN (B-W)

TWO-FACTOR MIXED DESIGN[3]

Follow through the steps below.

- *Have you read the* Limitations and Exceptions *section?*

- *Are all of these assumptions met for your study? If not, do not continue.*

- *Verbal equivalents for all notation are presented in the* Computations *section below (see also* Note on Notation, *p. 133).*

- *If you are experienced with computations for the analysis of variance, go directly to Summary Table 8.6 and proceed in your usual way. If you require additional guidance, begin with Section B,* The Example *(referring back to earlier sections as indicated).*

- *If your design has more (or fewer) levels, simply expand (or contract) the design by continuing any particular step until all levels or level combinations have been accounted for.*

A. (See Table 8.6.)

---

[3] This design is also called a "split-plot design."

Table 8.6. Summary Table: Between-Within ANOVA

Source	df	SS = Sum of Squares	MS = Mean Squares	F
Between-Subjects	$s-1$	$SS_S = \sum_{s=1}^{s} T_S^2/n_S - G^2/N$	(not computed)	
A	$a-1$	$SS_A = \sum_{A=1}^{a} T_A^2/n_A - G^2/N$	$MS_A = SS_A/(a-1)$	$F_A = MS_A/MS_{E:BS}$
Error: Between-Subjects	$s-a$	$SS_{E:BS} = SS_S - SS_A$	$MS_{E:BS} = SS_{E:BS}/(s-a)$	
Within-Subjects	$s(b-1)$	$SS_{WS} = SS_T - SS_S$	(not computed)	
B	$b-1$	$SS_B = \sum_{B=1}^{b} T_B^2/n_B - G^2/N$	$MS_B = SS_B/(b-1)$	$F_B = MS_B/MS_{E:WS}$
AB	$(a-1) \cdot (b-1)$	$SS_{AB} = \sum_{AB=1}^{ab} T_{AB}^2/n_{AB} - SS_A - SS_B - G^2/N$	$MS_{AB} = SS_{AB}/(a-1)(b-1)$	$F_{AB} = MS_{AB}/MS_{E:WS}$
Error: Within-Subjects	$(b-1) \cdot (s-a)$	$SS_{E:WS} = SS_{WS} - SS_B - SS_{AB}$	$MS_{E:WS} = SS_{E:WS}/(b-1)(s-a)$	
Total	$N-1$	$SS_T = \sum X^2 - G^2/N$		

B. THE EXAMPLE

A psychologist investigated the rate of acquisition in different species of rodents that learned a complex Hampton Court maze. Ten albino rats and ten Mongolian gerbils ran the maze repeatedly, with each subject receiving a total of 30 trials. The interval between trials was set at one minute. Completion of the maze resulted in food reward. The investigator measured the total number of blind alleys entered by each animal on each run. The data were grouped into three blocks of ten trials each. The mean number of blind alleys entered by each of the 20 subjects was recorded for each block of ten trials in a 2 X 3 data table.

C. RECORDING THE DATA

In all the steps below, A and B indicate the names of the two factors; a and b indicate the number of levels of each factor.

1. Record the data in an a X b table (see Table 8.7). Our present design is a two-factor design with two levels of the first (the between-subjects) and three levels of the second (the within-subjects) variable (2 X 3 table). *Please note that Factor B is the within-subjects variable and that Factor A is the between-subjects variable.* This is essential for following the computational steps. For computational convenience, each individual row of numbers should comprise the data for only one subject. This means that Factor B should be the column head, as in Table 8.7.

D. INITIAL STEPS

Many of these steps are spelled out in greater detail for a somewhat comparable analysis in the *Initial Steps* to the two-way within-subjects ANOVA, p. 175.

1. Begin a Summary Table. Include the column heads for this table exactly as they are indicated in Summary Table 8.6. Include the row heads (the items listed under "Source"), being sure that you use the appropriate *name of the factors* as well as the A and B labels. The Summary Table for the example appears in Table 8.8. Complete the table as you proceed through the computations below. Each critical step is marked by the notation: "Record in Summary Table."

2. Compute the degrees of freedom (df), using the formulas in the df column of Table 8.6, and record them in your Summary Table. See Table 8.8 for the df for this example.

Table 8.7. Raw-Data Table: Between-Within ANOVA (2 × 3)

			Factor B Trials (Within)*			
Factor A Species (Between)*			Block 1 (B_1)	Block 2 (B_2)	Block 3 (B_3)	Subjects
Rats (A_1)	S_1		12	5	3	$T_{S_1} = 20$ ($n_S = 3$)
	S_2		8	6	2	$T_{S_2} = 16$
	S_3		9	5	2	$T_{S_3} = 16$
	S_4		11	6	3	$T_{S_4} = 20$
	S_5		6 $T_{A_1B_1} = 102$	7 $T_{A_1B_2} = 59$	1 $T_{A_1B_3} = 22$	$T_{S_5} = 14$
	S_6		9	6	2	$T_{S_6} = 17$ $T_{A_1} = 183$
	S_7		15 ($n_{AB} = 10$)	7 ($n_{AB} = 10$)	4 ($n_{AB} = 10$)	$T_{S_7} = 26$ ($n_A = 30$)
	S_8		9	5	2	$T_{S_8} = 16$
	S_9		16	7	2	$T_{S_9} = 25$
	S_{10}		7	5	1	$T_{S_{10}} = 13$
Gerbils (A_2)	S_{11}		9	4	2	$T_{S_{11}} = 15$ ($n_S = 3$)
	S_{12}		9	3	2	$T_{S_{12}} = 14$
	S_{13}		6	2	2	$T_{S_{13}} = 10$
	S_{14}		10	2	2	$T_{S_{14}} = 14$
	S_{15}		6 $T_{A_2B_1} = 82$	2 $T_{A_2B_2} = 30$	2 $T_{A_2B_3} = 19$	$T_{S_{15}} = 10$ $T_{A_2} = 131$
	S_{16}		8	2	1	$T_{S_{16}} = 11$
	S_{17}		10 ($n_{AB} = 10$)	3 ($n_{AB} = 10$)	3 ($n_{AB} = 10$)	$T_{S_{17}} = 16$ ($n_A = 30$)
	S_{18}		12	6	2	$T_{S_{18}} = 20$
	S_{19}		7	3	2	$T_{S_{19}} = 12$
	S_{20}		5	3	1	$T_{S_{20}} = 9$
			$T_{B_1} = 184$ ($n_B = 20$)	$T_{B_2} = 89$ ($n_B = 20$)	$T_{B_3} = 41$ ($n_B = 20$)	($N = 60$) $G = 314$

*It is absolutely essential in the Raw-Data Table that Factor B be your within variable (that is, every subject took B_1, B_2, and so on). Conversely, it is essential that Factor A be your between variable (that is, a different group of subjects took A_1, A_2, and so on). Check closely because an error at this point is catastrophic.

Table 8.8. Summary Table: Between-Within ANOVA

Source	df	SS	MS	F
Between-Subjects	19	137.4	—	
Species (A)	1	45.066	45.066	8.79
Error: Between-Subjects	18	92.334	5.129	
Within-Subjects	40	627.333	—	
Trials (B)	2	529.633	264.816	118.81
Species × Trials (A × B)	2	17.434	8.717	3.91
Error: Within-Subjects	36	80.266	2.229	
Total	59	764.733		

3. There are 18 steps in this analysis. Since later computations refer back to steps by number, you will save time if you list the products from each step by number as you proceed. In addition, you can locate and correct errors in your computations more easily if you have recorded the intermediate steps systematically.

E. COMPUTATIONS

SUPERCHECK: No negative sums of squares (SS) may be obtained in the ANOVA. If you obtain a negative SS at any point, go back and check the preceding steps. Never go on if you have obtained a negative SS.

SUPERCHECK: *Each step* from 1 to 4 requires that *every score* be counted (in one category or another) *every time*. Be sure that you systematically count all scores. A check on each of these steps may be made by summing all the totals for any new step. *The resulting value must always be G.* (See *Check on Computations*, p. 134.)

1. a. Find $T_{A_1 B_1}, T_{A_1 B_2}, T_{A_1 B_3}, T_{A_2 B_1}, T_{A_2 B_2}, T_{A_2 B_3}$.

Sum the scores in each AB cell of the raw-data table. For convenience, record these totals in the cells of the table.

$T_{A_1 B_1} = 102, T_{A_1 B_2} = 59, T_{A_1 B_3} = 22, T_{A_2 B_1} = 82, T_{A_2 B_2} = 30,$
$T_{A_2 B_3} = 19$ (see cell totals, Table 8.7)

b. Find n_{AB}. [$n_{AB} = s/a$]

Determine the number of scores on which each of the AB totals is based. Record n_{AB}, in parentheses, immediately below the appropriate total.

$$\boxed{n_{AB} = 10 \qquad [10 = 20/2]}$$

2. a. Find T_{A_1}, T_{A_2}.

Sum the scores in each row of the raw-data table. Record these totals in the row margins of the table.

$$\boxed{T_{A_1} = 183, T_{A_2} = 131 \text{ (see row totals)}}$$

b. Find n_A. [$n_A = b \times n_{AB}$]

Determine the number of scores on which each of the two A-factor totals is based. For each total, record n_A, in parentheses, immediately below the appropriate total.

$$\boxed{n_A = 30 \qquad [30 = 3 \times 10]}$$

3. a. Find T_{B_1}, T_{B_2}, T_{B_3}.

Sum the scores in each column of the raw-data table. Record these totals in the column margins of the table.

$$\boxed{T_{B_1} = 184, T_{B_2} = 89, T_{B_3} = 41 \text{ (see column totals)}}$$

b. Find n_B. [$n_B = a \times n_{AB}$]

Determine the number of scores on which each of the two B-factor totals is based. For each total, record n_B, in parentheses, immediately below the appropriate total.

$$\boxed{n_B = 20 \qquad [20 = 2 \times 10]}$$

4. a. Find T_{S_1}, T_{S_2}, T_{S_3}, T_{S_4}, and so on.

Sum all the scores for each individual subject (by summing across the levels of B). Record these scores along the row margin of the table.

$$T_{S_1} = 20, T_{S_2} = 16, T_{S_3} = 16, T_{S_4} = 20 \ldots T_{S_{18}} = 20,$$
$$T_{S_{19}} = 12, T_{S_{20}} = 9 \text{ (see row margins)}$$

b. Find n_S. $[n_S = b]$

Determine the number of scores on which each subject's total is based. For each total, record n_S, in parentheses, beside the appropriate total.

$$n_S = 3 \qquad [b = 3]$$

5. a. Find G.

Sum all the scores. Record the sum at the bottom of the raw-data table.

$$G = 314$$

CHECK: $T_{A_1} + T_{A_2} = T_{B_1} + T_{B_2} + T_{B_3} = T_{S_1} + T_{S_2} + T_{S_3}$
$$+ T_{S_4} \ldots T_{S_8} = G$$

$$183 + 131 = 184 + 89 + 41 = 20 + 16 + 16$$
$$+ 20 + \cdots + 20 + 12 + 9 = 314$$

b. Find N.

Count the total number of scores in the raw-data table. Record N immediately below G, in parentheses.

$$N = 60$$

CHECK: $N = a \times b \times n_{AB}$

$$60 = 2 \times 3 \times 10. \qquad \text{OK!}$$

6. Find G^2/N.

Square G and divide the obtained value by N.

$$314^2/60 = 98,596/60 = 1643.27$$

7. Find ΣX^2

Square each individual's score in the $A \times B$ raw-data table and sum the squared values.

$$12^2 + 5^2 + 3^2 + 8^2 \ldots + 2^2 + 5^2 + 3^2 + 1^2 = 2408$$

8. Find $SS_{\text{Total}} = \Sigma X^2 - G^2/N$.

Subtract: #7 − #6. Record in Summary Table.

$$2,408.00 - 1,643.27 = 764.73$$

CHECK: This value must be positive. G^2/N must be smaller than or equal to ΣX^2. If G^2/N is larger than ΣX^2, you have made an error.

$$\text{The value is positive, } G^2/N < \Sigma X^2. \qquad \text{OK!}$$

9. Find $SS_S = \sum_{S=1}^{s} T_S^2/n_S - G^2/N$.

a. Square each of the values from #4a, sum the squared values, and divide by the total number of scores for each subject, n_S (#4b).

$$20^2 + 16^2 + 16^2 + 20^2 + \cdots + 20^2 + 12^2 + 9^2 = 5342$$
$$5342/3 = 1780.67$$

b. From this value subtract #6. Record in Summary Table.

$$\boxed{1780.67 \; - \; 1643.27 \; = \; 137.40}$$

10. Find $SS_A = \overset{a}{\underset{A=1}{\Sigma}} T_A^2 / n_A - G^2/N$.

   a. Square each of the values from # 2a, sum the squared values, and divide by n_A (# 2b).

$$\boxed{\begin{array}{l} 183^2 \; + \; 131^2 \; = \; 50{,}650.00 \\ 50{,}650.00/30 \; = \; 1688.33 \end{array}}$$

   b. From this value subtract # 6. Record in Summary Table.

$$\boxed{1688.33 \; - \; 1643.27 \; = \; 45.06}$$

11. Find $SS_B = \overset{b}{\underset{B=1}{\Sigma}} T_B^2 / n_B - G^2/N$.

   a. Square each of the values from # 3a, sum the squared values, and divide by n_B (# 3b).

$$\boxed{\begin{array}{l} 184^2 \; + \; 89^2 \; + \; 41^2 \; = \; 43{,}458.00 \\ 43{,}458.00/20 \; = \; 2172.90 \end{array}}$$

   b. From this value subtract # 6. Record in Summary Table.

$$\boxed{2172.90 \; - \; 1643.27 \; = \; 529.63}$$

12. Find $SS_{AB} = \overset{ab}{\underset{AB=1}{\Sigma}} T_{AB}^2 / n_{AB} - SS_A - SS_B - G^2/N$.

   a. Square each of the values from # 1a, sum the squared values, and divide by n_{AB} (# 1b).

$$\boxed{\begin{array}{l} 102^2 \; + \; 82^2 \; + \; 59^2 \; + \; 30^2 \; + \; 22^2 \; + \; 19^2 \; = \; 22{,}354.00 \\ 22{,}354.00/10 \; = \; 2{,}235.40 \end{array}}$$

   b. From this value subtract # 10b, # 11b, and # 6. Record in Summary Table.

$$2235.40 - 45.06 - 529.63 - 1643.27 = 17.44$$

13. Find $SS_{WS} = SS_{\text{Total}} - SS_S$.

    Subtract: #8 − #9b. Record in Summary Table.

$$764.73 - 137.40 = 627.33$$

14. Find $SS_{E:BS} = SS_S - SS_A$.

    Subtract: #9b − #10b. Record in Summary Table.

$$137.40 - 45.06 = 92.34$$

15. Find $SS_{E:WS} = SS_{WS} - SS_B - SS_{AB}$.

    Subtract: #13 − #11b − #12b. Record in Summary Table.

$$627.33 - 529.63 - 17.44 = 80.26$$

16. Compute the five mean squares indicated in Summary Table 8.6 by dividing each sum of squares by the corresponding df. Record in Summary Table.

    a. $MS_A = \#10/(a-1) = 45.06/1 = 45.06$

    b. $MS_{E:BS} = \#14/(s-a) = 92.34/18 = 5.13$

    c. $MS_B = \#11/(b-1) = 529.63/2 = 264.82$

    d. $MS_{AB} = \#12/(a-1)(b-1) = 17.44/2 = 8.72$

    e. $MS_{E:WS} = \#15/(b-1)(s-a) = 80.26/36 = 2.23$

17. Find the three Fs indicated by dividing mean squares by the appropriate error term (mean-square error) as indicated in Summary Table 8.6. Record in Summary Table.

    a. $F_A = \#16a/\#16b = 45.06/5.13 = 8.79$

    b. $F_B = \#16c/\#16e = 264.82/2.23 = 118.81$

    c. $F_{AB} = \#16d/\#16e = 8.72/2.23 = 3.91$

18. Evaluate computed *F*s by consulting Appendix 3, using the appropriate *df*.

---

a. $df_1 = 1, df_2 = 18, \alpha = .05, F_{\text{Tabled}} = 4.41$

$df_1 = 2, df_2 = 36, \alpha = .05, F_{\text{Tabled}} = 3.26$

$df_1 = 2, df_2 = 36, \alpha = .05, F_{\text{Tabled}} = 3.26$

b. $8.79 > 4.41$. Difference between rats and gerbils is significant. Reject the null hypothesis.

$118.81 > 3.26$. Differences among trials are significant. Reject the null hypothesis.

$3.91 > 3.26$. The interaction between species and trials is significant. Reject the null hypothesis.

F. INTERPRETATION OF THE EXAMPLE

1. *Statistical conclusion.* See p. 136 for a general statement of the statistical conclusions.
2. *Researcher's conclusion.* The number of blind alleys entered decreases over time for both species; however, gerbils engage in fewer errors overall than rats. The decrease in the final block of trials was smaller for the gerbils than for the rats.
3. *Journal summary.* Though the number of blind-alley entries decreased significantly with time for both species, the gerbils ($\bar{X} = 43.66$) engaged in fewer errors overall than the rats ($\bar{X} = 61.00$), $F(1,18) = 8.79, p < .05$. The means for the three blocks of trials were: Block 1, 92.0; Block 2, 44.5; Block 3, 20.5. An analysis of variance performed on these data revealed significant differences among trial means, $F(2,36) = 118.81, p < .05$. [Note that a specific-comparison test would be necessary to show which means differed significantly from the others.] The significant Species × Trials interaction $F(2,36) = 3.91, p < .05$, suggests that the decrease in errors in the final block of trials was smaller for the gerbils than for the rats. This interaction is shown in Figure 8.1.

G. FURTHER PROCEDURES

1. Interpretation: see Chapter 9, *Introduction* and Section C.
2. Specific-comparison tests: see Chapter 10, *Introduction* and Section C.
3. Strength-of-association measure: see Chapter 11, *Introduction* and Section C.

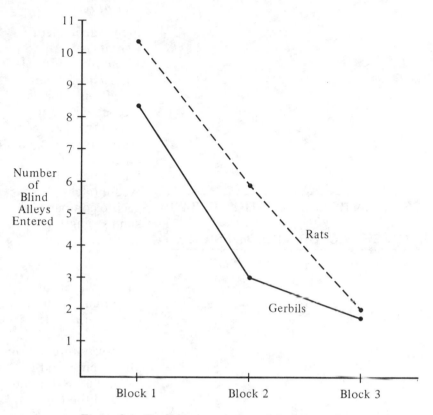

Figure 8.1. *Graphic presentation of the interaction.*

Description:

Three independent
   variables
Any number of levels
   of each variable
One within-subjects
   (repeated) factor
   AND two between-
   subjects (independent)
   factors
ns must be equal
Fixed model only
Specific comparisons
   and strength-of-
   association tests
   available
Tests for differences
   among levels of A;
   differences among
   levels of B; differences
   among levels of C;
   two-way interactions
   between A and B,
   A and C, and B and
   C; and the triple
   interaction A, B, and
   C

> **BETWEEN-BETWEEN-WITHIN (B-B-W)**
>
> **THREE-FACTOR MIXED DESIGN[4]**

Follow through the steps below.

- *Have you read the* Limitations and Exceptions *section?*

- *Are all of these assumptions met for your study? If not, do not continue.*

- *Verbal equivalents for all notation are presented in the* Computations *section below (see also* Note on Notation, *p. 133).*

- *If you are experienced with computation for the analysis of variance, go directly to Summary Table 8.9 and proceed in your usual way. If you require step-by-step guidance, begin with Section B,* Recording the Data.

- *For convenience in simplifying notation, we have used a design with a fixed set of levels. If your design has more (or fewer) levels, simply expand (or contract) the design by continuing any particular step until all levels or level combinations have been accounted for.*

A. (See Table 8.9)

[4]This design is also called a "split-plot design."

Table 8.9. Summary Table: Between-Between-Within ANOVA

Source	df	SS = Sum of Squares	MS = Mean Square	F
Between-Subjects	$s-1$	$SS_S = \sum_{S=1}^{s} T_S^2/n_S - G^2/N$	(not computed)	
A	$a-1$	$SS_A = \sum_{A=1}^{a} T_A^2/n_A - G^2/N$	$MS_A = SS_A/(a-1)$	$F_A = MS_A/MS_{E:BS}$
B	$b-1$	$SS_B = \sum_{B=1}^{b} T_B^2/n_B - G^2/N$	$MS_B = SS_B/(b-1)$	$F_B = MS_B/MS_{E:BS}$
AB	$(a-1)\cdot(b-1)$	$SS_{AB} = \sum_{AB=1}^{ab} T_{AB}^2/n_{AB} - SS_A$ $- SS_B - G^2/N$	$MS_{AB} = SS_{AB}/(a-1)(b-1)$	$F_{AB} = MS_{AB}/MS_{E:BS}$
Error: Between-Subjects	$s-ab$	$SS_{E:BS} = SS_S - SS_A - SS_B$ $- SS_{AB}$	$MS_{E:BS} = SS_{E:BS}/(s-ab)$	
Within-Subjects	$s(c-1)$	$SS_{WS} = SS_{\text{Total}} - SS_S$	(not computed)	
C	$c-1$	$SS_C = \sum_{C=1}^{c} T_C^2/n_C - G^2/N$	$MS_C = SS_C/(c-1)$	$F_C = MS_C/MS_{E:WS}$
AC	$(a-1)\cdot(c-1)$	$SS_{AC} = \sum_{AC=1}^{ac} T_{AC}^2/n_{AC} - SS_A$ $- SS_C - G^2/N$	$MS_{AC} = SS_{AC}/(a-1)(c-1)$	$F_{AC} = MS_{AC}/MS_{E:WS}$

Table 8.9 (continued)

Source	df	SS = Sum of Squares	MS = Mean Square	F
BC	$(b-1) \cdot$ $(c-1)$	$SS_{BC} = \sum\limits_{BC=1}^{bc} T^2_{BC}/n_{BC} - SS_B$ $- SS_C - G^2/N$	$MS_{BC} = SS_{BC}/(b-1)(c-1)$	$F_{BC} = MS_{BC}/MS_{E:ws}$
ABC	$(a-1) \cdot$ $(b-1) \cdot$ $(c-1)$	$SS_{ABC} = \sum\limits_{ABC=1}^{abc} T^2_{ABC}/n_{ABC} \cdot$ $- SS_A - SS_B - SS_C$ $- SS_{AB} - SS_{AC}$ $- SS_{BC} - G^2/N$	$MS_{ABC} = SS_{ABC}/(a-1) \cdot$ $(b-1)(c-1)$	$F_{ABC} = MS_{ABC}/MS_{E:ws}$
Error: Within-Subjects	$(c-1) \cdot$ $(s-ab)$	$SS_{E:ws} = SS_{ws} - SS_C - SS_{AC}$ $- SS_{BC} - SS_{ABC}$	$MS_{E:ws} = SS_{E:ws}/(c-1) \cdot$ $(s-ab)$	
Total	$N-1$	$SS_{Total} = \sum X^2 - G^2/N$		

B. RECORDING THE DATA

In all the steps below, A, B, and C indicate the names of the three factors; a, b, and c indicate the number of levels of each factor.

1. Record the data in an $a \times b \times c$ table. (see Table 8.10). As a convenient case, we shall select for general coverage a three-factor design with two levels of each variable ($2 \times 2 \times 2$ table). *Please note that Factor C is the within-subjects variable and that Factors A and B are the between-subjects variables.* This distinction is essential! For computational convenience, each individual row of numbers should constitute the data for only one subject. Thus Factor C should be the column head, as in Table 8.10.

C. INITIAL STEPS

Many of these steps are spelled out in greater detail for a somewhat comparable analysis in the *Initial Steps* to the two-way within-subjects ANOVA, p. 175.

1. Construct an $A \times C$ table to contain totals for each of the AC cells. To do this, ignore the third variable (B) and sum together all values for any particular AC combination (see Step #3). Be sure that A is the row factor and C is the column factor, just as in the original raw-data table (see Table 8.11a).

2. Construct a $B \times C$ table to contain totals for each of the BC cells. To do this, ignore the third variable (A) and sum together all values for any particular BC combination (see step #4). Be sure B is the row factor and C is the column factor, just as in the original raw-data table (see Table 8.11b).

3. Begin a Summary Table. Include the column heads for this table exactly as they are indicated in Summary Table 8.9. Include the row heads (the items listed under "Source"), being sure that you use the appropriate *name of the factors* as well as the labels A, B, and so on. Complete the table as you proceed through the computations below. Each critical step is marked by the notation: "Record in Summary Table."

4. Compute the degrees of freedom (df), using the formulas in the df column of Table 8.9, and record them in your Summary Table.

5. There are 26 steps in this analysis. Since later computations refer back to steps by number, you will save time if you list the products from each step by number as you proceed. In addition, you can locate and correct errors in your computations more easily if you have recorded the intermediate steps systematically.

Table 8.10. Raw-Data Table: Between-Between-Within ANOVA (2 × 2 × 2)

A Factor	B Factor		*C Factor (Within)*: C_1		C_2			
A_1	B_1	S_1	X_1		X_1		T_{S_1} (n_S)	
		S_2	X_2		X_2		T_{S_2} (n_S)	
		S_3	X_3	$T_{A_1B_1C_1}$	X_3	$T_{A_1B_1C_2}$	T_{S_3} (n_S)	$T_{A_1B_1}$
		.	.	(n_{ABC})	.	(n_{ABC})	.	(n_{AB})
		.	.		.		.	
		.	.		.		.	
		S_6	$X_{n_{ABC}}$		$X_{n_{ABC}}$		T_{S_6} (n_S)	
	B_2	S_7	X_1		X_1		T_{S_7} (n_S)	
		S_8	X_2		X_2		T_{S_8} (n_S)	
		S_9	X_3	$T_{A_1B_2C_1}$	X_3	$T_{A_1B_2C_2}$	T_{S_9} (n_S)	$T_{A_1B_2}$
		.	.		.		.	
		.	.	(n_{ABC})	.	(n_{ABC})	.	(n_{AB})
		.	.		.		.	
			$X_{n_{ABC}}$		$X_{n_{ABC}}$			
A_2	B_1	S_{13}	X_1		X_1		$T_{S_{13}}$ (n_S)	
			X_2		X_2			
			X_3		X_3			
			.	$T_{A_2B_1C_1}$	.	$T_{A_2B_1C_2}$	.	$T_{A_2B_1}$
			.	(n_{ABC})	.	(n_{ABC})	.	(n_{AB})
			.		.		.	
			$X_{n_{ABC}}$		$X_{n_{ABC}}$			
	B_2	S_{19}	X_1		X_1		$T_{S_{19}}$ (n_S)	
			X_2		X_2			
			X_3	$T_{A_2B_2C_1}$	X_3	$T_{A_2B_2C_2}$		$T_{A_2B_2}$
			.		.		.	
			.	(n_{ABC})	.	(n_{ABC})	.	(n_{AB})
			.		.		.	
		S_s	$X_{n_{ABC}}$		$X_{n_{ABC}}$		T_{S_s} (n_S)	
			T_{C_1} (n_C)		T_{C_2} (n_C)		G (N)	

Table 8.11

a

$A \times C$ Table: Between-Between-Within

	C_1	C_2	
A_1	$T_{A_1 C_1}$ (n_{AC})	$T_{A_1 C_2}$ (n_{AC})	T_{A_1} (n_A)
A_2	$T_{A_2 C_1}$ (n_{AC})	$T_{A_2 C_2}$ (n_{AC})	T_{A_2} (n_A)
	T_{C_1} (n_C)	T_{C_2} (n_C)	G (N)

b

$B \times C$ Table: Between-Between-Within

	C_1	C_2	
B_1	$T_{B_1 C_1}$ (n_{BC})	$T_{B_1 C_2}$ (n_{BC})	T_{B_1} (n_B)
B_2	$T_{B_2 C_1}$ (n_{BC})	$T_{B_2 C_2}$ (n_{BC})	T_{B_2} (n_B)
	T_{C_1} (n_C)	T_{C_2} (n_C)	G (N)

D. COMPUTATIONS

SUPERCHECK: *No negative sums of squares (SS) may be obtained in the ANOVA.* If you obtain a negative *SS* at any point, go back and check the preceding steps. Never go on if you have obtained a negative *SS*.

SUPERCHECK: *Each step* from 1 to 8 requires that *every score* be counted (in one category or another) *every time*. Be sure you systematically count all scores. A check on each of these steps can be made by summing all the totals for any step. *The resulting value must always be G* (see *Checks on Computations*, p. 134).

1. a. Find $T_{A_1 B_1 C_1}$, $T_{A_1 B_1 C_2}$, $T_{A_1 B_2 C_1}$, $T_{A_1 B_2 C_2}$, $T_{A_2 B_1 C_1}$, $T_{A_2 B_1 C_2}$, $T_{A_2 B_2 C_1}$, $T_{A_2 B_2 C_2}$.

Sum the scores in each ABC cell of the raw-data table. For your convenience, record these totals in the appropriate cells of the table.

b. Find n_{ABC}. [$n_{ABC} = s/(a \times b)$]

Determine the number of scores on which each of the ABC totals is based. Record n_{ABC}, in parentheses, in the table immediately below the appropriate total.

2. a. Find $T_{A_1 B_1}, T_{A_1 B_2}, T_{A_2 B_1}, T_{A_2 B_2}$.

Sum the scores in each of the four rows of the raw-data table. The total for each row should correspond to one of the four AB combinations; that is, sum all the scores, ignoring C. Record these totals in the row margins of the table.

b. Find n_{AB}. [$n_{AB} = c \times n_{ABC}$]

Determine the number of scores on which each of the AB totals is based. For each total, record n_{AB}, in parentheses, in the table immediately below the appropriate total.

3. a. Find $T_{A_1 C_1}, T_{A_1 C_2}, T_{A_2 C_1}, T_{A_2 C_2}$.

Sum the scores for each of the four AC combinations; that is, sum all the scores, ignoring B. Record these values in the cells of the $A \times C$ table.

b. Find n_{AC}. [$n_{AC} = b \times n_{ABC}$]

Determine the number of scores on which each of the AC totals is based. For each total, record n_{AC}, in parentheses, in the table immediately below the appropriate total.

4. a. Find $T_{B_1 C_1}, T_{B_1 C_2}, T_{B_2 C_1}, T_{B_2 C_2}$.

Sum all the scores for each of the four BC combinations; that is, sum all of these scores, ignoring A. Record these values in the cells of the $B \times C$ table.

b. Find n_{BC}. [$n_{BC} = a \times n_{ABC}$]

Determine the number of scores on which each of the BC totals is based. For each total, record n_{BC}, in parentheses, immediately below the appropriate total.

5. a. Find T_{A_1}, T_{A_2}.

Sum the scores in each row of the $A \times C$ table. For convenience, record these totals in the row margins of the $A \times C$ table.

b. Find n_A. [$n_A = b \times c \times n_{ABC}$]

Determine the number of scores on which each of the two A-factor totals is based. For each total record n_A, in parentheses, immediately below the appropriate total.

6. a. Find T_{B_1}, T_{B_2}.

Sum the scores in each row of the $B \times C$ table. For convenience, record these totals in the row margins of the $B \times C$ table.

b. Find n_B. $[n_B = a \times c \times n_{ABC}]$

Determine the number of scores on which each of the two B-factor totals is based. For each total, record n_B, in parentheses, immediately below the appropriate total.

7. a. Find T_{C_1}, T_{C_2}.

Sum the scores in each column of the $B \times C$ table. (These values must equal the summed scores in each column of the $A \times C$ table.) For convenience, record these totals in the column margins of each table.

b. Find n_C. $[n_C = a \times b \times n_{ABC}]$

Determine the number of scores on which each of the two C-factor totals is based. For each total, record n_C, in parentheses, immediately below the appropriate total in each table.

8. a. Find T_{S_1}, T_{S_2}, T_{S_3}, T_{S_4}, and so on.

Sum all the scores for each subject by summing the values in each row of the raw-data table.

b. Find n_S. $[n_S = c]$

Determine the number of scores on which each of the S totals is based. For each total, record n_S, in parentheses, beside the total in the row margins of the raw-data table.

9. a. Find G.

Sum all the scores. Record at the bottom of the raw-data table.

CHECK: $T_{A_1} + T_{A_2} = T_{B_1} + T_{B_2} = T_{C_1} + T_{C_2} = G$

$$T_{S_1} + T_{S_2} + T_{S_3} + T_{S_4} + \cdots + T_{S_S} = G$$

b. Find N.

Count the total number of scores in the raw-data table. Record N immediately below G, in parentheses, at the bottom of the raw-data table.

CHECK: $N = a \times b \times c \times n_{ABC}$

10. Find G^2/N.

Square G and divide the obtained value by N.

11. Find ΣX^2.

Square each individual score in the raw-data table. Then sum the squared values.

12. Find $SS_{\text{Total}} = \Sigma X^2 - G^2/N$.

Subtract: #11 − #10. Record in Summary Table.

CHECK: This value must be positive. G^2/N must be smaller than or equal to ΣX^2. If G^2/N is larger than ΣX^2, you have made an error.

13. Find $SS_S = \sum\limits_{S=1}^{s} T_S^2/n_S - G^2/N$.

   a. Square each of the values from #8a, sum the squared values, and divide by the number of scores for each subject, n_S (#8b).

   b. From this value subtract #10. Record in Summary Table.

14. Find $SS_A = \sum\limits_{A=1}^{a} T_A^2/n_A - G^2/N$.

   a. Square each of the values from #5a, sum the squared values, and divide by n_A (#5b).

   b. From this value subtract #10. Record in Summary Table.

15. Find $SS_B = \sum\limits_{B=1}^{b} T_B^2/n_B - G^2/N$.

   a. Square each of the values from #6a, sum the squared values, and divide by n_B (#6b).

   b. From this value subtract #10. Record in Summary Table.

16. Find $SS_C = \sum\limits_{C=1}^{c} T_C^2/n_C - G^2/N$.

   a. Square each of the values from #7a, sum the squared values, and divide by n_C (#7b).

   b. From this value subtract #10. Record in Summary Table.

17. Find $SS_{AB} = \sum\limits_{AB=1}^{ab} T_{AB}^2/n_{AB} - SS_A - SS_B - G^2/N$.

   a. Square each of the values from #2a, sum the squared values, and divide by n_{AB} (#2b).

   b. From this value subtract #14b, #15b, and #10. Record in Summary Table.

18. Find $SS_{AC} = \sum\limits_{AC=1}^{ac} T_{AC}^2/n_{AC} - SS_A - SS_C - G^2/N$.

   a. Square each of the values from #3a, sum the squared values, and divide by n_{AC} (#3b).

   b. From this value subtract #14b, #16b, and #10. Record in Summary Table.

19. Find $SS_{BC} = \sum\limits_{BC=1}^{bc} T_{BC}^2/n_{BC} - SS_B - SS_C - G^2/N$.

    a. Square each of the values from # 4a, sum the squared values, and divide by n_{BC} (# 4b).

    b. From this value subtract # 15b, # 16b, and # 10. Record in Summary Table.

20. Find $SS_{ABC} = \sum\limits_{ABC=1}^{abc} T_{ABC}^2/n_{ABC} - SS_A - SS_B - SS_C - SS_{AB} - SS_{AC} - SS_{BC} - G^2/N$.

    a. Square each of the values from # 1a, sum the squared values, and divide by n_{ABC} (# 1b).

    b. From this value subtract # 14b, # 15b, # 16b, # 17b, # 18b, # 19b, and # 10. Record in Summary Table.

21. Find $SS_{WS} = SS_{Total} - SS_S$.

    Subtract # 12 − # 13b. Record in Summary Table.

22. Find $SS_{E:BS} = SS_S - SS_A - SS_B - SS_{AB}$.

    Subtract: # 13b − # 14b − # 15b − # 17b. Record in Summary Table.

23. Find $SS_{E:WS} = SS_{WS} - SS_C - SS_{AC} - SS_{BC} - SS_{ABC}$.

    Subtract: # 21 − # 16b − # 18b − # 19b − # 20b. Record in Summary Table.

24. Compute the nine mean squares indicated in Summary Table 8.9 by dividing each sum of squares by its corresponding *df*. Record in Summary Table.

25. Find the seven *F*s by dividing mean squares by the appropriate error term (mean-square error) as indicated in Summary Table 8.9. Record in Summary Table.

26. Evaluate computed *F*s by consulting Appendix 3, using the appropriate *df* combination.

E. FURTHER PROCEDURES

1. Interpretation: see Chapter 9, *Introduction* and Section C.
2. Specific-comparison tests: see Chapter 10, *Introduction* and Section C.
3. Strength-of-association measure: see Chapter 11, *Introduction* and Section C.

Description:

Three independent
variables
Any number of levels of
each variable
Two within-subjects
(repeated) factors AND
one between-subjects
(independent) factor
Fixed model only
ns must be equal
Specific comparisons and
strength-of-association
tests available
Tests for differences
among the A factors;
differences among the
B factors; differences
among the C factors;
two-way interactions
between A and B, A
and C, and B and C;
and the triple interaction
among A, B, and C

BETWEEN-WITHIN-WITHIN (B-W-W)

THREE-FACTOR MIXED DESIGN[5]

Follow through the steps below.

- *Have you read the* Limitations and Exceptions *section?*

- *Are all of these assumptions met for your study? If not, do not continue.*

- *Verbal equivalents for all notation are presented in the* Computations *section below (see also* Note on Notation, *p. 133).*

- *If you are experienced with computation for the analysis of variance, go directly to Summary Table 8.12 and proceed in your usual way. If you require step-by-step guidance, begin with Section B,* Recording the Data.

- *For convenience in simplifying notation, we have used designs with a fixed set of levels. If your design has more (or fewer) levels, simply expand (or contract) the design by continuing any particular step until all levels or level combinations have been accounted for.*

[5]This design is also called a "split-plot design."

Table 8.12. Summary Table: Between-Within-Within ($a \times b \times c$)

Source	df	$SS = Sum\ of\ Squares$	$MS = Mean\ Square$	F
Between-Subjects	$s-1$	$SS_S = \sum\limits_{S=1}^{s} T_S^2/n_S - G^2/N$	(not computed)	
A	$a-1$	$SS_A = \sum\limits_{A=1}^{a} T_A^2/n_A - G^2/N$	$MS_A = SS_A/(a-1)$	$F_A = MS_A/MS_{E\,:\,BS}$
Error: Between-Subjects	$s-a$	$SS_{E\,:\,BS} = SS_S - SS_A$	$MS_{E\,:\,BS} = SS_{E\,:\,BS}/(s-a)$	
Within-Subjects	$s(bc-1)$	$SS_{WS} = SS_{Total} - S_S$	(not computed)	
B	$b-1$	$SS_B = \sum\limits_{B=1}^{b} T_B^2/n_B - G^2/N$	$MS_B = SS_B/(b-1)$	$F_B = MS_B/MS_{E_1\,:\,WS}$
C	$c-1$	$SS_C = \sum\limits_{C=1}^{c} T_C^2/n_C - G^2/N$	$MS_C = SS_C/(c-1)$	$F_C = MS_C/MS_{E_2\,:\,WS}$
AB	$(a-1)\cdot(b-1)$	$SS_{AB} = \sum\limits_{AB=1}^{ab} T_{AB}^2/n_{AB} - SS_A - SS_B - G^2/N$	$MS_{AB} = SS_{AB}/(a-1)(b-1)$	$F_{AB} = MS_{AB}/MS_{E_1\,:\,WS}$
AC	$(a-1)\cdot(c-1)$	$SS_{AC} = \sum\limits_{AC=1}^{ac} T_{AC}^2/n_{AC} - SS_A - SS_C - G^2/N$	$MS_{AC} = SS_{AC}/(a-1)(c-1)$	$F_{AC} = MS_{AC}/MS_{E_2\,:\,WS}$

Table 8.12 (continued)

Source	df	SS = Sum of Squares	MS = Mean Square	F
BC	$(b-1)\cdot$ $(c-1)$	$SS_{BC} = \sum_{BC=1}^{bc} T_{BC}^2/n_{BC} - SS_B - SS_C - G^2/N$	$MS_{BC} = SS_{BC}/(b-1)(c-1)$	$F_{BC} = MS_{BC}/MS_{E_3 : ws}$
ABC	$(a-1)\cdot$ $(b-1)\cdot$ $(c-1)$	$SS_{ABC} = \sum_{ABC=1}^{abc} T_{ABC}^2/n_{ABC} - SS_A - SS_B - SS_C - SS_{AB} - SS_{AC} - SS_{BC} - G^2/N$	$MS_{ABC} = SS_{ABC}/(a-1)\cdot$ $(b-1)(c-1)$	$F_{ABC} = MS_{ABC}/MS_{E_3 : ws}$
Error: Within-Subjects	$(bc-1)\cdot$ $(s-a)$	$SS_{E:ws} = SS_{ws} - \sum_{ABC=1}^{abc} T_{ABC}^2/n_{ABC} + G^2/N + SS_A$		
Error$_1$: Within-Subjects	$(b-1)\cdot$ $(s-a)$	$SS_{E_1:ws} = \sum_{ABS=1}^{abs} T_{ABS}^2/n_{ABS} - G^2/N - SS_S - SS_B - SS_{AB}$	$MS_{E_1:ws} = SS_{E_1:ws}/(b-1)\cdot$ $(s-a)$	

Table 8.12 (continued)

Source	df	SS = Sum of Squares	MS = Mean Square	F
Error₂: Within-Subjects	$(c-1) \cdot (s-a)$	$SS_{E_2:WS} = \sum_{ACS=1}^{acs} T_{ACS}^2/n_{ACS}$ $- G^2/N - SS_S$ $- SS_C - SS_{AC}$	$MS_{E_2:WS} = SS_{E_2:WS}/(c-1) \cdot$ $(s-a)$	
Error₃: Within-Subjects	$(b-1) \cdot$ $(c-1) \cdot$ $(s-a)$	$SS_{E_3:WS} = SS_{E:WS} - SS_{E_1:WS} - SS_{E_2:WS}$	$MS_{E_3:WS} = SS_{E_3:WS}/(b-1) \cdot$ $(c-1)(s-a)$	
Total	$N-1$	$SS_{Total} = \sum X^2 - G^2/N$		

A. (See Table 8.12.)

B. RECORDING THE DATA

In all the steps below, A, B, and C indicate the names of the three factors; a, b, and c indicate the number of levels of each factor.

1. Record the data in an $a \times b \times c$ raw-data table (see Table 8.13). As a convenient case, we'll select for general coverage a three-factor design with two levels of the first two factors and three levels of the third ($2 \times 2 \times 3$ table). *Please note that factors B and C are both within-subjects variables and that factor A is the between-subjects variable.* This distinction is essential! For computational convenience, each individual row of numbers should constitute data for only one subject.

C. INITIAL STEPS

Many of these steps are spelled out in greater detail for a somewhat comparable analysis in the *Initial Steps* to the two-way within-subjects ANOVA, p. 175.

1. Construct an $A \times B$ table to contain totals for each of the AB cells. To do this, ignore the third variable, C, and sum together all values for any particular AB combination (see Step #3). Be sure A is the row factor and B the column factor, just as in the original raw-data table (see Table 8.14a).

2. Construct an $A \times C$ table to contain totals for each of the AC cells. To do this, ignore the third variable, B, and sum together all values for any particular AC combination (see Step #4). Be sure A is the row factor and C the column factor, just as in the original raw-data table (see Table 8.14b).

3. Construct an $A \times B \times S$ table to contain totals for each subject in each of the AB combinations. Again, ignore the third variable, C, and for each subject sum together all values for a particular AB combination. The cells for each subject should contain that subject's total for the AB combination indicated (see Step #9). Be sure A is the row variable and B the column variable, as in the original raw-data table (see Table 8.14c).

4. Construct an $A \times C \times S$ table to contain totals for each subject in each of the AC combinations. Again, ignore the third variable, B, and for each subject sum together all values for a particular AC combination. The cells for each subject should contain that subject's total for the AC combination indicated (see Step #10). Be sure A is the row variable and C the column variable, as in the original raw-data table (see Table 8.14d).

Table 8.13. Raw-Data Table: Between-Within-Within (2 × 2 × 3)

	Factor B B_1 — Factor C			Factor B B_2 — Factor C			
	C_1	C_2	C_3	C_1	C_2	C_3	
S_1	X_1	X_1	X_1	X_1	X_1	X_1	$T_{S_1}\,(n_S)$
S_2	X_2	X_2	X_2	X_2	X_2	X_2	$T_{S_2}\,(n_S)$
S_3	$X_3\ T_{A_1B_1C_1}$	$X_3\ T_{A_1B_1C_2}$	$X_3\ T_{A_1B_1C_3}$	$X_3\ T_{A_1B_2C_1}$	$X_3\ T_{A_1B_2C_2}$	$X_3\ T_{A_1B_2C_3}$	$T_{S_3}\,(n_S)$
S_4	X_4	X_4	X_4	X_4	X_4	X_4	$T_{S_4}\,(n_S)$
.	(n_{ABC})	(n_{ABC})	(n_{ABC})	(n_{ABC})	(n_{ABC})	(n_{ABC})	.
.	.	.	.	.	.	.	.
S_{10}	$X_{n_{ABC}}$	$X_{n_{ABC}}$	$X_{n_{ABC}}$	$X_{n_{ABC}}$	$X_{n_{ABC}}$	$X_{n_{ABC}}$	$T_{S_{10}}\,(n_S)$
S_{11}	X_1	X_1	X_1	X_1	X_1	X_1	$T_{S_{11}}\,(n_S)$
S_{12}	X_2	X_2	X_2	X_2	X_2	X_2	$T_{S_{12}}\,(n_S)$
S_{13}	$X_3\ T_{A_2B_1C_1}$	$X_3\ T_{A_2B_1C_2}$	$X_3\ T_{A_2B_1C_3}$	$X_3\ T_{A_2B_2C_1}$	$X_3\ T_{A_2B_2C_2}$	$X_3\ T_{A_2B_2C_3}$	$T_{S_{13}}\,(n_S)$
.	(n_{ABC})	(n_{ABC})	(n_{ABC})	(n_{ABC})	(n_{ABC})	(n_{ABC})	.
.	.	.	.	.	.	.	.
S_S	$X_{n_{ABC}}$	$X_{n_{ABC}}$	$X_{n_{ABC}}$	$X_{n_{ABC}}$	$X_{n_{ABC}}$	$X_{n_{ABC}}$	$T_{S_S}\,(n_S)$
	$T_{B_1C_1}\,(n_{BC})$	$T_{B_1C_2}\,(n_{BC})$	$T_{B_1C_3}\,(n_{BC})$	$T_{B_2C_1}\,(n_{BC})$	$T_{B_2C_2}\,(n_{BC})$	$T_{B_2C_3}\,(n_{BC})$	G / N

A_1 (rows S_1–S_{10}), A_2 (rows S_{11}–S_S)

Factor A (Between)

Table 8.14

a

$A \times B$ Table: Between-Within-Within

	B_1	B_2	
A_1	$T_{A_1 B_1}$ (n_{AB})	$T_{A_1 B_2}$ (n_{AB})	T_{A_1} (n_A)
A_2	$T_{A_2 B_1}$ (n_{AB})	$T_{A_2 B_2}$ (n_{AB})	T_{A_2} (n_A)
	T_{B_1} (n_B)	T_{B_2} (n_B)	G (N)

b

$A \times C$ Table: Between-Within-Within

	C_1	C_2	C_3	
A_1	$T_{A_1 C_1}$ (n_{AC})	$T_{A_1 C_2}$ (n_{AC})	$T_{A_1 C_3}$ (n_{AC})	T_{A_1} (n_A)
A_2	$T_{A_2 C_1}$ (n_{AC})	$T_{A_2 C_2}$ (n_{AC})	$T_{A_2 C_3}$ (n_{AC})	T_{A_2} (n_A)
	T_{C_1} (n_C)	T_{C_2} (n_C)	T_{C_3} (n_C)	G (N)

c

$A \times B \times S$ Table: Between-Within-Within

		B_1	B_2	
	S_1	$T_{S_1 A_1 B_1}$ (n)	$T_{S_1 A_1 B_2}$ (n)	T_{S_1} (n_S)
	S_2	$T_{S_2 A_1 B_1}$ (n)	$T_{S_2 A_1 B_2}$ (n)	T_{S_2} (n_S)
	S_3	$T_{S_3 A_1 B_1}$ (n)	$T_{S_3 A_1 B_2}$ (n)	T_{S_3} (n_S)
A_1	.	.	.	.
	.	.	.	.
	.	.	.	.
	S_{10}	$T_{S_{10} A_1 B_1}$ (n)	$T_{S_{10} A_1 B_2}$ (n)	$T_{S_{10}}$ (n_S)
	S_{11}	$T_{S_{11} A_2 B_1}$ (n)	$T_{S_{11} A_2 B_2}$ (n)	$T_{S_{11}}$ (n_S)
	S_{12}	$T_{S_{12} A_2 B_1}$ (n)	$T_{S_{12} A_2 B_2}$ (n)	$T_{S_{12}}$ (n_S)
	S_{13}	$T_{S_{13} A_2 B_1}$ (n)	$T_{S_{13} A_2 B_2}$ (n)	$T_{S_{13}}$ (n_S)
A_2	.	.	.	.
	.	.	.	.
	.	.	.	.
	S_S	$T_{S_s A_2 B_1}$ (n)	$T_{S_s A_2 B_2}$ (n)	T_{S_s} (n_S)
		T_{B_1} (n_B)	T_{B_2} (n_B)	G (N)

d

$A \times C \times S$ Table: Between-Within-Within

		C_1	C_2	C_3	
	S_1	$T_{S_1 A_1 C_1}$ (n)	$T_{S_1 A_1 C_2}$ (n)	$T_{S_1 A_1 C_3}$ (n)	T_{S_1} (n_S)
	S_2	$T_{S_2 A_1 C_1}$ (n)	$T_{S_2 A_1 C_2}$ (n)	$T_{S_2 A_1 C_3}$ (n)	T_{S_2} (n_S)
	S_3	$T_{S_3 A_1 C_1}$ (n)	$T_{S_3 A_1 C_2}$ (n)	$T_{S_3 A_1 C_3}$ (n)	T_{S_3} (n_S)
A_1	.	.	.	.	.
	.	.	.	.	.
	.	.	.	.	.
	S_{10}	$T_{S_{10} A_1 C_1}$ (n)	$T_{S_{10} A_1 C_2}$ (n)	$T_{S_{10} A_1 C_3}$ (n)	$T_{S_{10}}$ (n_S)
	S_{11}	$T_{S_{11} A_2 C_1}$ (n)	$T_{S_{11} A_2 C_2}$ (n)	$T_{S_{11} A_2 C_3}$ (n)	$T_{S_{11}}$ (n_S)
	S_{12}	$T_{S_{12} A_2 C_1}$ (n)	$T_{S_{12} A_2 C_2}$ (n)	$T_{S_{12} A_2 C_3}$ (n)	$T_{S_{12}}$ (n_S)
A_2	S_{13}	$T_{S_{13} A_2 C_1}$ (n)	$T_{S_{13} A_2 C_2}$ (n)	$T_{S_{13} A_2 C_3}$ (n)	$T_{S_{13}}$ (n_S)
	.	.	.	.	.
	.	.	.	.	.
	.	.	.	.	.
	S_S	$T_{S_s A_2 C_1}$ (n)	$T_{S_s A_2 C_2}$ (n)	$T_{S_s A_2 C_3}$ (n)	T_{S_s} (n_S)
		T_{C_1} (n_C)	T_{C_2} (n_C)	T_{C_3} (n_C)	G (N)

5. Begin a Summary Table. Include the column heads for this table exactly as they are indicated in Summary Table 8.12. Include the row heads (the items listed under "Source"), being sure that you use the appropriate *names of the factors* as well as the labels A, B, and so on. Complete the table as you proceed through the computations below. Each critical step is marked by the notation: "Record in Summary Table."

6. Compute the degrees of freedom (df), using the formulas in the df column of Table 8.12, and record them in your Summary Table.

7. There are 31 steps in this analysis. Since later computations refer back to steps by number, you will save time if you list the products from each step by number as you proceed. In addition, you can locate and correct errors in your computations more easily if you have recorded the intermediate steps systematically.

D. COMPUTATIONS

SUPERCHECK: *No negative sums of squares (SS) may be obtained in the ANOVA.* If you obtain a negative SS at any point, go back and check the preceding steps. Never go on if you have obtained a negative SS.

SUPERCHECK: *Each step* from 1 to 10 requires that *every score* be counted (in one category or another) *every time*. Be sure that you systematically count all scores. A check on each of these steps can be made by summing all the totals for any step. *The resulting value must always be G* (see *Checks on Computations*, p. 134).

1. a. Find $T_{A_1 B_1 C_1}$, $T_{A_1 B_1 C_2}$, $T_{A_1 B_1 C_3}$, $T_{A_1 B_2 C_1}$, $T_{A_1 B_2 C_2}$, $T_{A_1 B_2 C_3}$, $T_{A_2 B_1 C_1}$, $T_{A_2 B_1 C_2}$, $T_{A_2 B_1 C_3}$, $T_{A_2 B_2 C_1}$, $T_{A_2 B_2 C_2}$, $T_{A_2 B_2 C_3}$.
   Sum the scores in each ABC cell of the raw-data table. For your convenience, record these values in the table.
   b. Find n_{ABC}. [$n_{ABC} = s/a$]
   Determine the number of scores on which each of the ABC totals is based. Record this, n_{ABC}, in parentheses, in the table immediately below the appropriate total.

2. a. Find $T_{B_1 C_1}$, $T_{B_1 C_2}$, $T_{B_1 C_3}$, $T_{B_2 C_1}$, $T_{B_2 C_2}$, $T_{B_2 C_3}$.
   Sum the scores in each of the six columns of the raw-data table. The total for each column corresponds to one of the six BC combinations.

b. Find n_{BC}. [$n_{BC} = a \times n_{ABC}$]

Determine the number of scores on which each of the BC totals is based. For each total, record n_{BC}, in parentheses, in the column margins immediately below the total.

3. a. Find $T_{A_1B_1}$, $T_{A_1B_2}$, $T_{A_2B_1}$, $T_{A_2B_2}$.

Sum all the scores for each of the four AB conditions; that is, sum all AB scores, ignoring C. Record these values in the cells of the $A \times B$ table.

b. Find n_{AB}. [$n_{AB} = c \times n_{ABC}$]

Determine the number of scores on which each of the AB totals is based. For each total, record n_{AB}, in parentheses, immediately below the appropriate total.

4. a. Find $T_{A_1C_1}$, $T_{A_1C_2}$, $T_{A_1C_3}$, $T_{A_2C_1}$, $T_{A_2C_2}$, $T_{A_2C_3}$.

Sum all the scores for each of the six AC combinations; that is, sum all AC scores, ignoring B. Record these values in the cells of the $A \times C$ table.

b. Find n_{AC}. [$n_{AC} = b \times n_{ABC}$]

Determine the number of scores on which each of the AC totals is based. For each total, record n_{AC}, in parentheses, immediately below the appropriate total.

5. a. Find T_{A_1}, T_{A_2}.

Sum the scores in each row of the $A \times B$ (or $A \times C$) table.

b. Find n_A. [$n_A = b \times c \times n_{ABC}$]

Determine the number of scores on which each of the two A-factor totals is based. For each total, record n_A, in parentheses, immediately below the appropriate total.

6. a. Find T_{B_1}, T_{B_2}.

Sum the scores in each column of the $A \times B$ table.

b. Find n_B. [$n_B = a \times c \times n_{ABC}$]

Determine the number of scores on which each of the two B-factor totals is based. For each total, record n_B, in parentheses, immediately below the appropriate total.

7. a. Find T_{C_1}, T_{C_2}, T_{C_3}.

Sum the scores in each column of the $A \times C$ table.

b. Find n_C. [$n_C = a \times b \times n_{ABC}$]

Determine the number of scores on which each of the three C-factor totals is based. For each total, record n_C, in parentheses, immediately below the appropriate total.

8. a. Find T_{S_1}, T_{S_2}, T_{S_3}, and so on.

Sum all the scores for each subject (by summing each row of the raw-data table).

b. Find n_S. [$n_S = b \times c$]

Determine the number of scores on which each of the S totals is based. For each total, record n_S, in parentheses, beside the total in the row margins of the raw-data table.

9. a. Find $T_{A_1 B_1 S_1}$, $T_{A_1 B_1 S_2}$, $T_{A_1 B_1 S_3}$, and so on.

For each subject, sum separately the three scores (for C_1, C_2, and C_3) in each AB combination. In this case, you will obtain four scores for each subject. Record these totals in the cells of your $A \times B \times S$ table.

b. Find n_{ABS}. [$n_{ABS} = c$]

Determine the number of scores on which each of the ABS totals is based. For each total, record n_{ABS}, in parentheses, in the cells of the $A \times B \times S$ table.

10. a. Find $T_{A_1 C_1 S_1}$, $T_{A_1 C_2 S_1}$, $T_{A_1 C_3 S_1}$, $T_{A_2 C_1 S_1}$, $T_{A_2 C_2 S_1}$, $T_{A_2 C_3 S_1}$, and so on.

For each subject, sum separately the two (for B_1 and B_2) scores in each AC combination. In this case, you will obtain six scores for each subject. Record these totals in the cells of your $A \times C \times S$ table.

b. Find n_{ACS}. [$n_{ACS} = b$]

Determine the number of scores on which each of the ACS totals is based. For each total, record n_{ACS}, in parentheses, beside the totals in the cells of the $A \times C \times S$ table.

11. a. Find G.

Sum all the scores. Record them at the bottom of the raw-data table.

CHECK: $T_{A_1} + T_{A_2} = T_{B_1} + T_{B_2} = T_{C_1} + T_{C_2} + T_{C_3} = G$.

$$T_{S_1} + T_{S_2} + T_{S_3} + T_{S_4} + \cdots + T_{S_S} = G.$$

b. Find N. [$N = a \times b \times c \times n_{ABC}$]

Count the total number of scores in the raw-data table. Record N immediately below G, in parentheses.

12. Find G^2/N.

Square G and divide the obtained value by N.

13. Find ΣX^2.

Square each individual score in the raw-data table and sum the squared values.

14. Find $SS_{\text{Total}} = \Sigma X^2 - G^2/N$.

Subtract: #13 − #12. Record in Summary Table.

CHECK: This value must be positive. G^2/N must be smaller than or equal to ΣX^2. If G^2/N is larger than ΣX^2, you have made an error.

15. Find $SS_S = \sum\limits_{S=1}^{s} T_S^2/n_S - G^2/N$.

    a. Square each of the values from # 8a, sum the squared values, and divide by the number of scores for each subject, n_S (# 8b).

    b. From this value subtract # 12. Record in Summary Table.

16. Find $SS_A = \sum\limits_{A=1}^{a} T_A^2/n_A - G^2/N$.

    a. Square each of the values from # 5a, sum the squared values, and divide by n_A (# 5b).

    b. From this value subtract # 12. Record in Summary Table.

17. Find $SS_B = \sum\limits_{B=1}^{b} T_B^2/n_B - G^2/N$.

    a. Square each of the values from # 6a, sum the squared values, and divide by n_B (# 6b).

    b. From this value subtract # 12. Record in Summary Table.

18. Find $SS_C = \sum\limits_{C=1}^{c} T_C^2/n_C - G^2/N$.

    a. Square each of the values from # 7a, sum the squared values, and divide by n_C (# 7b).

    b. From this value subtract # 12. Record in Summary Table.

19. Find $SS_{AB} = \sum\limits_{AB=1}^{ab} T_{AB}^2/n_{AB} - SS_A - SS_B - G^2/N$.

    a. Square each of the values from # 3a, sum the squared values, and divide by n_{AB} (# 3b).

    b. From this value subtract # 16b, # 17b, and # 12. Record in Summary Table.

20. Find $SS_{AC} = \sum\limits_{AC=1}^{ac} T_{AC}^2/n_{AC} - SS_A - SS_C - G^2/N$.

    a. Square each of the values from # 4a, sum the squared values, and divide by n_{AC} (# 4b).

    b. From this value subtract # 16b, # 18b, and # 12. Record in Summary Table.

21. Find $SS_{BC} = \sum\limits_{BC=1}^{bc} T_{BC}^2/n_{BC} - SS_B - SS_C - G^2/N$.

    a. Square each of the values from # 2a, sum the squared values, and divide by n_{BC} (# 2b).

    b. From this value subtract # 17b, # 18b, and # 12. Record in Summary Table.

22. Find $SS_{ABC} = \sum\limits_{ABC=1}^{abc} T^2_{ABC}/n_{ABC} - SS_A - SS_B - SS_C - SS_{AB}$
    $- SS_{AC} - SS_{BC} - G^2/N$.
    a. Square each of the values from # 1a, sum the squared values, and divide by n_{ABC} (# 1b). *Save this value!*
    b. From this value subtract # 16b, # 17b, # 18b, # 19b, # 20b, # 21b, and # 12. Record in Summary Table.
23. Find $SS_{WS} = SS_{\text{Total}} - SS_S$.
    Subtract: # 14 − # 15b. Record in Summary Table.
24. Find $SS_{\text{Error}:BS} = SS_S - SS_A$.
    Subtract: # 15b − # 16b. Record in Summary Table.
25. Find $SS_{\text{Error}:WS} = SS_{WS} - \sum\limits_{ABC=1}^{abc} T^2_{ABC}/n_{ABC} + G^2/N + SS_A$.
    Subtract: # 23 − # 22a + # 12 + # 16b.

    NOTE: The last two values are added.

    Record in Summary Table.
26. Find $SS_{\text{Error}_1:WS} = \sum\limits_{ABS=1}^{abs} T^2_{ABS}/n_{ABS} - G^2/N - SS_S - SS_B$
    $- SS_{AB}$.
    a. Square each of the values from # 9a, sum the squared values, and divide by n_{ABS} (# 9b).
    b. Subtract: # 26a − # 12 − # 15b − # 17b − # 19b. Record in Summary Table.
27. Find $SS_{\text{Error}_2:WS} = \sum\limits_{ACS=1}^{acs} T^2_{ACS}/n_{ACS} - G^2/N - SS_S - SS_C$
    $- SS_{AC}$.
    a. Square each of the values from # 10a, sum the squared values, and divide by n_{ACS} (# 10b).
    b. Subtract: # 27a − # 12 − # 15b − # 18b − # 20b. Record in Summary Table.
28. Find $SS_{\text{Error}_3:WS} = SS_{E:WS} - SS_{E_1:WS} - SS_{E_2:WS}$.
    Subtract: # 25 − # 26 − # 27. Record in Summary Table.
29. Compute the 11 mean squares indicated in Summary Table 8.12 by dividing each sum of squares by its corresponding *df*. Record in Summary Table.
30. Find the seven *F*s by dividing the mean squares by the appropriate error term (mean-square error) as indicated in Summary Table 8.12. Record in Summary Table.
31. Evaluate computed *F*s by consulting Appendix 3, using the appropriate *df*.

D. FURTHER PROCEDURES

1. Interpretation: see Chapter 9, *Introduction* and Section C.
2. Specific-comparison tests: see Chapter 10, *Introduction* and Section C.
3. Strength-of-association measure: see Chapter 11, *Introduction* and Section C.

CHAPTER NINE

INTERPRETATION

The results of a statistical test of significance[1] permit you to say that there are or are not relationships between your independent and dependent variables. (You may prefer to say that there are or are not differences between or among conditions). The researcher who discovers relationships or differences must understand their meaning for the particular research problem being studied. The word *interpretation* refers to the problem of understanding and reporting the results of a particular study. The interpretation may be very simple or very complex, depending on the design of the study and the relationship among the variables.

The problem of interpreting research results is largely ignored in statistics books. This omission is not surprising, since exact rules useful for interpreting research findings are difficult to specify. Although our goal for this book was to present clear rules applicable to any research design, this task proved too formidable. We have tried, though, in a modest way, to present some general principles that may help you interpret your findings.

If the statistical test indicates nonsignificance (that there is no basis for rejecting the null hypothesis), then the interpretation is only that there is no relationship (or that there are no differences). It is generally of little interest to find that two variables are *not* related, and studies with such findings are rarely published except when there is a good theoretical reason to believe that the variables are related or when previous research indicates that a relationship does exist between them. There is a basic problem, however, in trying to demonstrate that there are no differences between groups. You can never be certain whether there are no real differences or whether your procedures simply are not sufficiently precise to show differences that do exist.

If the statistical test indicates significance (the null hypothesis is rejected), you know that there is a relationship between your

[1]It should be emphasized that the words "significant" or "significance" are used throughout this book in the statistical sense of "statistically significant finding" rather than in the more general sense of "meaningful" or "important."

independent and dependent variables. Knowing that a relationship exists, however, tells you nothing about the nature of that relationship. You must carefully inspect your data to determine the direction of the relationship and its meaning in the context of your particular research problem. In some cases, inspection of the data will indicate the exact nature of the relationship you have found. In other cases, further statistical procedures may be necessary. In general, the simpler the research design, the easier it is to determine the nature of the relationship by inspection. When there are two levels of a single independent variable, inspection alone, following your test of significance, is always sufficient. When the number of levels or the number of independent variables increases, specific-comparison tests may be necessary (see Chapter 10).

Research prediction, the alternative hypothesis, and the null hypothesis. Inexperienced researchers sometimes encounter additional difficulty when interpretation includes confirmation or disconfirmation of a research prediction or research hypothesis. The differences among the research prediction, the null hypothesis, and the alternative hypothesis have been treated in Chapter 4, but here we shall discuss their relationship to interpretation.

There are several null hypotheses (see footnote, p. 43), but we'll consider only the simplest of these: that the two or more treatment means (or other appropriate measures) are equal. The alternative hypothesis for this null hypothesis is that the means are not equal. This alternative is not directional; in other words, for the simple case in which two means are involved, the alternative is true both when Mean 1 is larger than Mean 2 and when Mean 2 is larger than Mean 1. For example, your research hypothesis is that women will perform better than men on a verbal-learning task. It is possible, however, that the men may perform better than the women. You establish the null hypothesis that there will be no differences between the means of these groups. Your alternative hypothesis is that the means will be different (in either direction).

Thus in this rather typical case the research hypothesis not only assumes a difference but specifies the direction in which that difference will occur. In general, if you do not reject your null hypothesis, you must say that your experimental prediction was not supported by your findings. On the other hand, if you reject your null hypothesis and accept the two-tailed alternative hypothesis, you may or may not be able to state that the research prediction was supported by your data. If your research hypothesis was merely that the groups differed from one another, rejection of the overall null hypothesis automatically indicates support[2] for the research prediction. However, most research predictions specify expected direction or an order to the differences between or among groups. In these cases, the differences must be in the predicted direction or in the predicted order before you can claim support for the research prediction. The significant results may well be in the direction opposite from the predicted direction. Such results clearly do not offer support for the research prediction. Before you may assert that the results of the study support or fail to support a *directional* research hypothesis, you must inspect the data and (sometimes) make specific-comparison tests. If your research design is complex and there are multiple research hypotheses, or if the hypotheses are themselves complex, it is even more important that you pay particular attention to the direction of the differences in the data.

In the pages that follow, we assume in every case that the null hypothesis being tested is the same null hypothesis (null hypothesis: general form) presented with the particular analysis in the earlier chapters (5, 6, 7, or 8). In all cases, rejection of (or failure to reject) this null hypothesis provides the basis for our interpretation. In addition, in each case we give the *research interpretation*, not the statistical interpretation, of the study. That is, we provide the type of interpretation that would be written in a journal article. Writers of scientific communications often trust that they will be understood by other scientists to be referring to populations when they make general statements, and they usually make no explicit mention of the population. Always remember, however, that interpreting statistically significant findings provides information about the population from which the sample is drawn, not simply about the sample. Thus when we say, "the proportions of people in each category differ" or "the two groups were significantly different from one another," we really mean, "in the population from which these subjects were drawn, the proportions of people falling into the categories are different," or "in the populations from which these scores were drawn, the mean scores are different."

It is sometimes difficult to make the statements of findings clear and appropriately general. One convention followed by scientific journals is to state things that are true for the population (general truths) in the present tense and to write most of the remainder of an article in the past tense. Although populations are rarely mentioned in scientific writing, the conscientious researcher always gives careful thought to the population to which he can reasonably generalize his

---

[2]Avoid saying that your findings *prove* your research hypothesis. The nature of proof is complicated, but "proof" generally is not involved in psychological research.

findings and specifies this population as clearly as possible. (See p. 50 for more discussion of populations).

For each statistical test presented in this book, we have listed a series of steps for interpreting the data. These steps provide a *summary guideline* that is particularly useful after you have some experience in interpreting data. Your first reading, therefore, should include a perusal of the *General Considerations and Examples* that follow the *Interpretational Steps*. Thereafter, you will need to consult the later sections only if you require additional clarification of the meaning of the interpretational steps.

Note that we have not provided interpretational steps for the case in which there are no statistically significant differences. Note also that, in cases in which there are more than one statistical outcome to be considered (all the two-or-more-variable designs), each combination of outcomes has been programmed separately, since a somewhat different combination of steps is required in each case.

Since different considerations arise depending on the kind of dependent variable involved, we look at the three types of dependent variables separately.

I. FREQUENCY DATA

A. ONE-WAY χ^2

Interpretational Steps
IF χ^2 IS SIGNIFICANT:
1. If $a = 2$:
    a. The obtained frequencies differ significantly from the expected frequencies.
    b. Compare each obtained frequency with its expected frequency. One obtained frequency is larger than its expected frequency and the other is smaller.
    c. There is a significantly greater number of subjects in the over-represented category and a significantly smaller number of subjects in the under-represented category than would be expected if the null hypothesis were true.
2. If $a > 2$:
    a. Overall, the obtained frequencies differ from the expected frequencies.
    IF A MORE SPECIFIC INTERPRETATION IS NEEDED:
    b. Compare each obtained frequency with its expected fre-

quency. One or more obtained frequencies are larger than their expected frequencies and one or more are smaller than their expected frequencies.

c. If it is not clear which obtained frequencies differ from their corresponding expected frequencies, an adaptation of Ryan's procedure (Chapter 10, p. 301) may be used to make comparisons between all pairs of obtained frequencies.

GENERAL CONSIDERATIONS AND EXAMPLES

The one-way χ^2 tests the null hypothesis that a set of observed frequencies do not differ from a set of expected frequencies—in other words, that the observed frequencies were drawn from a population of frequencies with the same proportions as the expected frequencies. Interpretations are generally rather simple with this design. For example, suppose that a poll has been conducted that asked people whether they were more in sympathy with the principles of the Democratic Party or of the Republican Party. Of the 500 people questioned, 300 expressed sympathy with Democratic principles and 200 with Republican principles. To test the null hypothesis that the

	Democrat	Republican	Total
Obtained	300	200	500
Expected	250	250	

number of people expressing sympathy with each party is equal, the expected frequency for each answer would be 250. The data indicate that Democrat answers are over-represented and Republican answers are under-represented. We would interpret a significant χ^2 to mean that, for the population from which these persons were drawn, there is greater sympathy for Democratic than for Republican party principles.

As the number of levels of independent variables increases, clear-cut interpretation becomes more difficult. For example, 200 randomly chosen urban dwellers were asked to indicate the meat they preferred: beef, lamb, pork, or chicken. For a null hypothesis that the meats would be chosen equally frequently, the expected frequency for each meat would be 50. Inspection of the data indicates that beef is chosen more often, lamb almost as often, chicken somewhat less often,

	Beef	*Lamb*	*Pork*	*Chicken*	*Total*
Obtained	92	51	17	40	200
Expected	50	50	50	50	

and pork considerably less often than would be expected under the null hypothesis. A significant χ^2 would indicate that the four meats were not chosen with equal frequency, but it would not tell which meats were chosen significantly more or less frequently than expected. However, in this type of study, the researcher would probably note only that, for this population, the four meats were not chosen with equal frequency and that beef was most preferred and pork least preferred. If you wished to make a more specific interpretation—that is, if you wished to compare each meat with each other meat—you might use an adaptation of Ryan's procedure (Chapter 10, p. 301) to make these comparisons. In a one-way χ^2, however, general interpretations are more usual.

B. TWO-WAY χ^2 ($2 \times 2, a \times b$)

Interpretational Steps
IF χ^2 IS SIGNIFICANT:
1. If $a = b = 2$:
   a. There is a significant association between A and B.
   b. Compute the proportion of A_1 at B_1 (frequency in $A_1 B_1$ cell divided by total frequency of A_1) and A_2 at B_1 (frequency in $A_2 B_1$ cell divided by total frequency of A_2).
   c. Compare $p_{A_1 B_1}$ with $p_{A_2 B_1}$. The larger proportion is significantly greater than the smaller proportion.[3]
2. If either (but not both) a or $b > 2$:
   a. Overall, there is a significant association between A and B.
   IF A MORE SPECIFIC INTERPRETATION IS NEEDED:
   b. Draw a graph showing the relationship of the two independent variables.
   c. If the graph does not make the relationship clear, use Ryan's procedure to make all possible 2×2 comparisons.
   d. Follow the procedures in Section 1, above, to interpret each significant 2×2.

---

[3] Alternatively, we may compare the proportion of A_1 at B_2 with the proportion of A_2 at B_2; or the proportion of B_1 at A_1 with the proportion of B_2 at A_1; or the proportion of B_1 at A_2 with the proportion of B_2 at A_2. However, all of these procedures do the same thing and will result in the same conclusions and interpretations.

3. If both a and $b > 2$:
   a. Overall, there is a significant association between A and B.
   IF A MORE SPECIFIC INTERPRETATION IS NEEDED:
   b. Draw a graph showing the relationship of the two independent variables.
   c. If the graph does not make the relationship clear, classifications may be combined or dropped out to produce a $2 \times b$ or an $a \times 2$ table, and Ryan's procedure may be used to make all possible 2×2 comparisons.
   d. Follow the procedures in Section 1, above, to interpret each significant 2×2.
   e. In some large tables with complex relationships between independent variables, precise interpretations may be difficult or impossible.

GENERAL CONSIDERATIONS AND EXAMPLES

Graphical vs tabular presentation. Some people find it easier to interpret χ^2 from tabular representation; others find graphical representation clearer. In this section, we have presented only tabular displays. If the contingency table for the χ^2 is confusing to you, you may graph the contingency table as you would graph a two-way interaction in the analysis of variance (see *Notes on Interactions*, p. 290). The only difference is that for χ^2 the values on the vertical axis of the graph indicate frequencies or proportions rather than scores.

The two-way χ^2 tests the null hypothesis that there is no relationship between two independent variables. The interpretation of the 2×2 χ^2 may be made from an inspection of the data. Suppose we have polled students from a small liberal-arts college to determine the number of freshmen and seniors who own or do not own cars. If χ^2 is significant, the researcher may conclude that, for this population, there is a relationship between academic class and car ownership. However, only an inspection of the two data columns successively can indicate the direction of the relationship; in this population, seniors

		A_1 Freshmen	A_2 Seniors	
B_1	Own Car	35	65	100
B_2	Do Not Own Car	120	42	162
		155	107	262

are more likely to own cars than are freshmen. (Note that this is a verbal statement of $P_{A_2 B_1} > P_{A_1 B_1}$.) An inspection of the two rows of data successively would lead to a similar interpretation of the data. It should be pointed out, however, that there are usually alternative ways of stating the outcome: for example, "freshmen are less likely to own cars . . . " ($P_{A_1 B_1} < P_{A_2 B_1}$), or even, "those who own cars are more likely to be seniors, while those who do not own cars are more likely to be freshmen" ($P_{A_1 B_2}$ and $P_{A_2 B_1} > P_{A_1 B_1}$ and $P_{A_2 B_2}$).

Data for a hypothetical study that adds sophomores, juniors, and graduate students to the sample provide more complex interpretational problems. A significant χ^2 again would indicate that, for this population, there was a relationship between academic class and

	Freshmen	*Sophomores*	*Juniors*	*Seniors*	*Graduates*	
Own Car	35	60	55	65	12	227
Do Not Own Car	120	49	50	42	41	302
	155	109	105	107	53	529

car ownership, but the relationship would not be at all clear from inspection. For example, fewer freshmen and graduate students than expected own cars, whereas more sophomores, juniors, and seniors than expected own cars. The difference between freshmen and sophomores is probably statistically significant, but the difference between sophomores and seniors probably is not. The researcher must use a specific-comparison test to determine the specific pattern of differences among classes. Ryan's procedure (Chapter 10, p. 301) will answer this type of question.

In many respects, the interpretation of a two-way χ^2 involves the same considerations as the interpretation of a two-way interaction in the analysis of variance; the reader is referred to *Note on Interactions*, p. 290. It may also be helpful to read the material on the interpretation of two-way analysis of variance, p. 279.

A test indicating a significant relationship by itself answers only a very general question about the data. In many cases, of course, the general answer is all that is needed. If the researcher asks more specific questions about the nature of the relationship, he must use further statistical procedures. As the research design becomes more complex, the need for these further procedures increases, although at times even they may be of little help. In a very large two-way χ^2 (for example, a 4 × 5 or 6 × 10 χ^2), the relationship may be so complex that even the use of specific-comparison tests may fail to make the relationship clear. Because of the problems of interpretation, the researcher should strive for simplicity in designing studies. Always avoid unnecessarily complex research designs. The additional factors or the additional levels may make interpretation of the results difficult, if not impossible.

C. THREE-WAY χ^2 (2 × 2 × 2, a × b × c)

Interpretational Steps
1. IF THE THREE-WAY χ^2 IS NOT SIGNIFICANT BUT AT LEAST ONE TWO-WAY χ^2 IS SIGNIFICANT:
   a. There is a significant association between the independent variables that enter into the significant two-way χ^2.
   b. Follow the interpretational rules given for two-way χ^2 (p. 266).
2. IF THE THREE-WAY χ^2 IS SIGNIFICANT BUT NO TWO-WAY χ^2 IS SIGNIFICANT:
   a. There is a significant association among the three independent variables.
   b. Inspect the tabular display of, or draw a graph of, the relationship between any two independent variables for each level of the third independent variable.
   c. Determine the relationship for each two-way χ^2.
   d. The associations for each two-way χ^2 are significantly different for different levels of the third independent variable.
3. IF THE THREE-WAY χ^2 IS SIGNIFICANT AND AT LEAST ONE TWO-WAY χ^2 IS ALSO SIGNIFICANT:
   a. Interpret the three-way χ^2 according to the steps given in #2, above.
   b. Interpret each two-way χ^2 following the steps for two-way χ^2. Remember to qualify the interpretation of each significant two-way χ^2, since the relationship between the two variables depends on the level of the third independent variable. The presence of the significant three-way χ^2 may make the significant two-way χ^2 relatively unimportant or meaningless.

GENERAL CONSIDERATIONS AND EXAMPLES

The three-way χ^2 tests four null hypotheses. Three of the null hypotheses state that there is no association between the independent variables taken two at a time. Interpretation of these χ^2 values follows the principles used for interpreting two-way χ^2, for that is what they are. For a 2 × 2 × 2 χ^2, the interpretations may be made easily by

inspecting corresponding *pairs* of columns (or rows) in the three
2 × 2 arrays. If there are more than two levels for any independent
variable, specific-comparison tests may be required.

The fourth null hypothesis states that the relationship between
any two independent variables is the same for the different levels of
the third independent variable. As an example, we present data for a
hypothetical study that examined the relationship among gender,
drug administration, and test performance. Our data are the frequency
of men and women passing or failing as a function of drug administration. In this first example, there is no significant three-way interaction and therefore the hypothesis of no association would be

	Male					*Female*		
	Pass	Fail				Pass	Fail	
Drug	12	28	40		Drug	13	26	39
No Drug	30	6	36		No Drug	32	7	39
	42	34	76			45	33	78

Two-Way Interaction Significant but Three-Way Interaction Not Significant

accepted. Since our two-way interaction between drug administration
and gender is significant, we conclude that administration of the drug
has the same deleterious effect for both sexes.

As a convenient comparison, let's examine a different set of
hypothetical frequencies for the same study. Assume that, in this
second example, only the three-way χ^2 is significant. Systematic

	Male					*Female*		
	Pass	Fail				Pass	Fail	
Drug	12	28	40		Drug	26	13	39
No Drug	30	6	36		No Drug	7	32	39
	42	34	76			33	45	78

Only Three-Way Interaction Significant

inspection of the rows (or columns) indicates that administration of
the drug was responsible for a decrease in performance among men
but an increase in performance among women. In other words, the
association between drug administration and performance was in one

direction for men and in the other direction for women. (It should be noted that the two-way χ^2 between performance and drugs would not be significant because the effects found for men would exactly cancel those found for women. However, *specific comparisons* would almost certainly indicate a significant relationship between these variables for men and women considered separately.) This interpretation was possible only from inspecting the data in tabular or graphical form. Although we chose to present the relationship between drug and performance for men and women in this study, we might have presented the relationship between gender and drug for pass and fail, or the relationship between gender and performance for each drug condition. (Exercise: reorder the data into these two alternative forms.) The interpretation does not depend on how the data are presented, but it is sometimes easier to see existing relationships in one configuration of the data than in another. For this reason, it is often a good idea to consider the relationship in all three configurations and see which makes the most sense to you. Again, the interpretation of a three-way χ^2 is very similar to the interpretation of a three-way interaction in an analysis of variance. It may be helpful to refer to the section on three-way analysis of variance in this chapter and to *Note on Interactions* for further explanation and examples.

A final possibility for this same study might be that the drug had no effect on women but a strongly deleterious effect on men. In this case, it is possible that both the three-way χ^2 and at least one two-way

	Men				Women		
	Pass	Fail			Pass	Fail	
Drug	12	28	40	Drug	18	21	39
No Drug	30	6	36	No Drug	18	21	39
	42	34	76		36	42	78

Three-Way and Two-Way Interactions Significant

χ^2 would be significant. Assume that the two-way χ^2 between drug administration and test performance is significant (that is, the data are collapsed across gender). We might interpret the two-way χ^2 to indicate that the drug was responsible for an overall decrease in performance. However, we recognize that the finding is misleading, since the three-way χ^2 indicates that all of the decreases can be accounted for by the performances of men. In this case, the meaning of the two-way χ^2 must be carefully qualified by the meaning of a significant three-way χ^2.

D. REPEATED-MEASURES χ^2 AND McNEMAR'S χ_c^2

Interpretational Steps

1. Repeated-measures χ^2

   IF χ^2 IS SIGNIFICANT:

   a. There is a significant association between the two independent variables. (Remember that scores for both independent variables represent measurements on the same subjects.)

   b. Follow the interpretational steps for the two-way χ^2.

2. McNemar's χ_c^2

   IF χ_c^2 IS SIGNIFICANT:

   a. The proportion in Category 1 on the first occasion is different from the proportion in Category 1 on the second occasion.

   b. Compare the number of subjects in Category 1 on the first occasion with the number in Category 1 on the second occasion.

   c. The larger of these two numbers is significantly greater than the smaller. More subjects changed in one direction than in the other.

GENERAL CONSIDERATIONS AND EXAMPLES

Although they are similar, the repeated-measures χ^2 and McNemar's χ_c^2 are used for different purposes. The repeated-measures χ^2 is generally used when the researcher wishes to determine whether two different measures on the same subject are related; in other words, it is a test of association. For example, subjects given two different personality tests might have their scores on each test dichotomized (above and below the median) or trichotomized (high, medium, and low). A significant repeated-measures χ^2 would indicate a relationship between performance on the two tests. The relationship could be positive (a person scores high on both tests or low on both tests) or negative (a person scores high on one test and low on the other). The direction of the relationship may be determined only by inspection of the data.

The McNemar χ_c^2 is generally used when the *same measurement* has been taken twice on the same subjects.[4] The researcher is interested not in a relationship between the two measures but in a possible differential change from the first measurement to the second. The McNemar statistic is significant when changes in one direction

---

[4]Remember that the two observations may involve matched pairs, litter mates, or twins as well as two observations on the same individual. Thus this test does not require that the two samples be independent. For simplicity, we consider the test only as it applies to two observations on the same individual, but an analogous interpretation is possible for the other cases.

exceed those in the other direction. When χ_c^2 is significant, inspect the data to determine the direction of the change.

II. ORDERED DATA

A. RANK-SUMS TEST, SIGN TEST

Interpretational Steps
1. Rank-Sums Test
   IF Z IS SIGNIFICANT:
   a. The two groups differ from one another.
   b. Determine which group has the higher sum of ranks and which has the lower.
   c. The group with the higher sum of ranks has significantly higher ranks than the group with the lower sum of ranks. Be sure that you understand the relationship between high and low ranks and the underlying behavioral dimension being measured.
2. Sign Test
   IF R IS SIGNIFICANT
   a. There is a significant change from Measurement 1 to Measurement 2.
   b. Determine whether there are more "+ changes" or more "− changes."
   c. If there are more "+ changes," there is a significant change from Measurement 1 to Measurement 2 in the + direction.
   d. If there are more "− changes," there is a significant change from Measurement 1 to Measurement 2 in the − direction.

GENERAL CONSIDERATIONS AND EXAMPLES

Since each of the above tests applies to research designs having only two levels of one independent variable, the interpretation may always be made simply by determining the larger sum of ranks or the larger number of changes. However, there are some special problems that apply to significance tests for ordinal data, since these techniques test differences between ranks rather than differences between numerical scores. You must remember that interpretations apply to ranks rather than to numerical scores (if any) that underlie the ranks. It is always possible for ranks to yield different results than would comparable analyses performed on numerical scores.

Remember, too, that for some tests, like the rank-sums test, ranks may be assigned in either ascending or descending order. Thus,

Table 9.1. Two Examples of an Experiment Relating Self-Esteem to Success or Failure

	Example A				Example B			
	Success		Failure		Success		Failure	
	Score	Rank	Score	Rank	Score	Rank	Score	Rank
	23	4	56	17	23	17	56	4
	18	1	58	18	18	20	58	3
	29	6	40	11	29	15	40	10
	31	7	61	19	31	14	61	2
	46	13	52	16	46	8	52	5
	22	3	36	10	22	18	36	11
	26	5	49	14	26	16	49	7
	19	2	50	15	19	19	50	6
	34	8	73	20	34	13	73	1
	42	12	35	9	42	9	35	12
	TOTAL	61		149		149		61

NOTE: In Example A, low ranks are assigned to low scores. In Example B, high ranks are assigned to low scores. In both examples, low scores indicate high self-esteem.

if one group has a high sum of ranks and the other a low sum of ranks, you must remember whether a high rank means a low or a high score, good or poor performance, and so on. Two hypothetical outcomes of a study with paper-and-pencil self-esteem scores for subjects exposed to a success experience or a failure experience illustrate the problems of interpreting the meanings of ranks. In Example A (see Table 9.1), the rank of 1 was assigned to the lowest score. In Example B, the rank of 1 was assigned to the highest score. Both examples utilize the same scores, and, in both, a low score indicates high self-esteem. A significant Z indicates that the two groups differ from one another. In both examples, the interpretation is that subjects who experienced success expressed greater self-esteem than subjects who experienced failure. In Example A, the researcher must remember both (1) that low sums of ranks were associated with low scores and (2) that low scores were associated with high self-esteem. In Example B, the researcher must remember both (1) that low sums of ranks were associated with high scores and (2) that high scores were associated with low self-esteem. Clearly, if the researcher becomes confused about the way ranks are related to the underlying behavior, he may make a completely reversed interpretation of his results.

B. KRUSKAL-WALLIS TEST, FRIEDMAN TEST

Interpretational Steps
1. Kruskal-Wallis Test
IF *H* IS SIGNIFICANT:
  a. Overall, the groups differ from one another.
  b. Inspect the sums of the ranks for each group to determine whether the differences among the groups are obvious.
  c. If the differences are not obvious, use Ryan's procedure to determine which pairs of groups differ. Be sure that you understand the relationship of high and low ranks to the underlying behavioral dimension being measured.
2. Friedman Test
IF χ_r^2 IS SIGNIFICANT:
  a. Overall, the sets of measurements on the same subjects differ from one another.
  b. Inspect the sums of the ranks for each set of measurements to determine whether the differences among the measurements are obvious.
  c. If the differences are not obvious, use Nemenyi's test (Chapter 10, p. 311) to determine which pairs of measurements differ. Be sure that you understand the relationship of high and low ranks to the underlying behavioral dimension being measured.

GENERAL CONSIDERATIONS AND EXAMPLES

In the Kruskal-Wallis and Friedman tests, as in the other ordinal tests, the researcher must keep in mind the relationship of the ranks to the underlying behavioral dimension (see *General Considerations and Examples*, p. 273). In addition, since both of these tests deal with data from more than two samples, there may be difficulties in pinpointing the exact meaning of a significant finding.

Let's assume that groups of rats have been tested in a maze under one of four levels of deprivation: 0, 4, 8, and 12 hours. The dependent variable is number of trials to some criterion of performance. The researcher has ranked these scores, and a Kruskal-Wallis test applied to the ranks indicates that the groups differ. The experimenter may conclude that level of deprivation is related to maze performance in some way. However, there are literally dozens of ways they can be related. To cite just a few: (1) all four groups may differ significantly from each other, (2) only the 12-hour group may differ from the others, (3) the 12-hour and 8-hour groups differ significantly from the 4-hour and 0-hour groups, but the 12-hour group does not differ significantly from the 8-hour group and the 4-hour group does not differ significantly from the 0-hour group, and so on. In some cases, you may make your interpretation by examining your sums of ranks; but in most cases you must use a specific-comparison test, such as Ryan's procedure (Chapter 10, p. 306) or Nemenyi's test (Chapter 10, p. 311). Some of the principles involved are easier to illustrate with score data; the researcher who has performed a Kruskal-Wallis test or a Friedman test should read the material on the one-way between-subjects analysis of variance and main effects in the next section. The same considerations apply to interpretation of the Kruskal-Wallis and Friedman tests.

III. SCORE DATA (ANALYSIS OF VARIANCE, *t* TEST)

A. ANALYSES WITH ONE INDEPENDENT VARIABLE

Between-Subjects Designs: one-way between-subjects ANOVA, *t* test
Within-Subjects Designs: a × *s, t* test for correlated means
Interpretational Steps
IF THE *F* RATIO OR *t* IS SIGNIFICANT:
1. Compute the means for each level of the independent variable.
2. a. If $a = 2$: the larger mean is significantly greater than the smaller mean.
   b. (1) If $a > 2$: the significant *F* indicates only that there are differences among the means, since there are several alternative ways in which means may differ among themselves.
   (2) A specific-comparison test must be performed to provide information on differences between particular means (see Chapter 10, p. 313).
3. For your summary: report the means, values of *F*, degrees of freedom (for both the numerator and denominator) of your *F* ratio, and your level of significance. Indicate the meaning of the finding in terms of your research hypothesis or problem.

GENERAL CONSIDERATIONS AND EXAMPLES

One-way between-subjects: two levels. The one-way between-subjects analysis of variance with two levels and the *t* test are used to analyze the results from the simplest type of score-data study, a between-subjects design involving two groups of subjects. The interpretation of such designs requires only that the direction of the difference of the means be observed. A significant *F* or *t* indicates that the greater mean is significantly larger than the smaller mean.

One-way between-subjects: a levels. When *a* is greater than 2, the interpretation is somewhat more complex because, whereas the analysis of variance answers a very general question about the data (for example, are there differences *among* the means of the groups?), the researcher is often interested in answering more specific questions (for example, is Group 1 different from Group 2?). If you wish to know whether there is a significant difference between any pair of means, specific-comparison tests must be performed. Let's return to the example from the section on the Kruskal-Wallis test. Groups of rats were tested in a maze under 0, 4, 8, or 12 hours of deprivation (see Table 9.2). The dependent variable is the number of trials to a

Table 9.2. Mean Number of Trials to Criterion for Groups of Rats Experiencing 0, 4, 8, or 12 Hours of Deprivation

		Hours of Deprivation			
	Examples	0	4	8	12
	A	22.4	21.8	10.6	10.3
Five	B	22.4	21.8	23.1	10.3
Hypothetical	C	22.4	11.1	10.9	10.8
Outcomes	D	26.8	19.6	13.2	8.1
	E	26.8	14.6	18.2	8.1

NOTE: The five examples represent different hypothetical outcomes for the experiment.

specified criterion. Let's consider several hypothetical examples of a number of possible outcomes from such an experiment. Assume that for each of the five examples listed, the obtained *F* ratio is significant.

In Example A, inspection of the data indicates that there are virtually identical means for the 0- and 4-hour groups and for the 8- and 12-hour groups. Specific-comparison tests would certainly show significant differences *between* these pairs of means. Since a large number of trials to criterion indicates slower learning, the appropriate interpretation is that the rats deprived 8 or 12 hours learned the maze significantly more quickly than the rats deprived 0 or 4 hours (or that the 0- and 4-hour groups required more trials to criterion than the 8- and 12-hour groups). In Example B, since the values for each group except the 12-hour one are virtually identical, specific-comparison tests would indicate significant differences between the 12-hour group and all the rest. In this case, the interpretation would be that rats deprived 12 hours learned the maze more quickly than either the 8-, 4-, or 0-hour deprived rats, and that the latter did not differ among themselves. In Example C, the rats deprived 0 hours differed from and were inferior to all of the others.

In Example D, speed of learning increased with increased deprivation (at least for the levels of deprivation studied). The researcher may make this statement of overall differences, may test for the significance of this trend,[5] or may test precisely for the pattern of differences among the groups. In the latter case, he would have to use a specific-comparison test to state which pairs of means differed. In Example E, there is a tendency for increased deprivation to be associated with faster learning, but there is a reversal at the 4- and 8-hour levels. In this case, the researcher would wish to use a specific-comparison test. Since one might expect a monotonic increase in performance with hours of deprivation, it would be important to know whether the 4- and 8-hour groups were significantly different from one another or whether the differences between them were due to chance.

Main effects. We have discussed in detail the interpretation of a one-way between-subjects analysis of variance with four levels. However, the same considerations apply to main effects from factorial experiments, regardless of whether they are between-subjects, within-subjects, or mixed designs. For example, in a 2 X 2 X 4 analysis of variance, the interpretation for the main effects of the first two independent variables may be made by inspection (since each has two levels). The main effect of the independent variable with four levels may be interpreted by applying the principles illustrated in our example. However, in factorial designs, the interpretation of main effects often is influenced by the presence of a significant interaction, a problem we take up when we consider two independent variables.

a X *s ANOVA (within-subjects) or t test for correlated means.* Although Table 7.12 for the *a* X *s* ANOVA resembles that of a two-way ANOVA without an error term[6] rather than a one-way ANOVA, the

---

[5] Such tests are beyond the scope of this text; however, they are covered very thoroughly by Edwards (1968), Chapter 14.

[6] In theory, it is a two-way analysis with only one subject per cell. The analysis partials out the variance attributable to the independent variable and the variance attributable to differences among subjects. However, a significant F ratio for subjects would mean only that people are different from one another, a fact that needs no further statistical demonstration. Thus "subjects" is not tested. There is no true (within cell variance) error term because, with only one score per cell, there are no degrees of freedom left over for error. However, the *a* X *s* interaction serves as an estimate of error for this analysis. To the extent that the treatment has a relatively constant effect on all subjects, the interaction of treatments with subjects is small and the possibility of a large F ratio for treatments is enhanced.

interpretation of the $a \times s$ analysis requires the same steps as a between-subjects one-way analysis. If there are two levels of the independent variable, a significant F ratio for treatments or a significant t indicates that one mean is significantly greater than the other. If there are more than two levels in the analysis of variance, specific comparisons may be needed to determine the locus of the differences. Refer to the one-way between-subjects (a levels) ANOVA for examples.

B. ANALYSES WITH TWO INDEPENDENT VARIABLES

Between-Subjects Designs: $2 \times 2, a \times b$
Within-Subjects Designs: $a \times b \times s$
Mixed Designs: $a \times b$ (B-W)

If you are unfamiliar with interactions or their interpretations, a review of *Notes on Interactions*, p. 290, before reading this section may be helpful.

Interpretational Steps

IF THE F RATIO FOR $A \times B$ INTERACTION IS NOT SIGNIFICANT BUT THE F RATIO FOR ONE OR BOTH MAIN EFFECTS IS SIGNIFICANT:

1. Compute means for each significant main effect. Consider each variable in turn (a refers to the number of levels of the variable under consideration).
2. a. If $a = 2$: the larger mean is significantly greater than the smaller mean.
   b. (1) If $a > 2$: the significant F indicates only that there are differences among the means, since there are several alternative ways in which the means may differ among themselves.
   (2) A specific-comparison test must be performed to provide information on differences between specific means (see Chapter 10, p. 313).
3. For your summary: report the means for the significant main effects, the value of F, the degrees of freedom, and the level of significance. Indicate the meaning of the finding in terms of your research hypothesis or problem.

IF THE F RATIO FOR $A \times B$ INTERACTION IS SIGNIFICANT AND NO MAIN EFFECT IS SIGNIFICANT:

1. Compute means for each cell of the $A \times B$ table.
2. Graph the means and visually inspect the meaning of the interaction (see Figure 9.1). If visual inspection does not clarify the meaning of the interaction, a Tukey's (a) test (see Chapter 10) may be performed to identify the locus of the significant difference.

3. For your summary: present means in tabular or graphic form. Report the value of F, the degrees of freedom, and the level of significance. Indicate the meaning of the interaction in terms of the research hypothesis or problem.

IF THE F RATIO FOR $A \times B$ INTERACTION IS SIGNIFICANT AND THE F RATIO FOR ONE OR BOTH MAIN EFFECTS IS ALSO SIGNIFICANT:

1. Compute means for each cell of the $A \times B$ table.
2. Graph the means and visually inspect the meaning of the interaction (see Figure 9.1). If visual inspection does not clarify the meaning of the interaction, Tukey's (a) test may be performed to identify the locus of the significant difference.
3. Compute means for the significant main effects. Consider each variable in turn (*a* refers to the number of levels of the variable under consideration).
4. a. If $a = 2$: the larger mean is significantly greater than the smaller mean.
   b. (1) If $a > 2$: the significant F indicates only that there are differences among the means, since there are several alternative ways in which the means may differ among themselves.
      (2) A specific-comparison test must be performed to provide information on differences between particular means (see Chapter 10, p. 313).
5. For your summary: report the means for the significant main effect, the value of F, the degrees of freedom, and the level of significance. Report the means for your interaction—graphically, if it adds clarity. Give the value of F, the degrees of freedom, and the level of significance. Indicate the meaning of the main effect in terms of your research problem, *then give the limitation of your statement introduced by the significant interaction.* Remember that the significant interaction always qualifies the kind of statement you make in interpreting the main effects.

GENERAL CONSIDERATIONS AND EXAMPLES

2 × 2 ANOVA (between-subjects). The 2 × 2 analysis of variance tests the effects of each of two factorially combined independent variables. The advantage of such a design is that, in addition to assessing the effects of each independent variable separately, it enables the researcher to investigate the interaction between the two variables. An interaction occurs whenever the effects of the levels of one independent variable depend on the specific levels of the second independent variable. We illustrate this point with the results of two hypothetical experiments (see Table 9.3).

Table 9.3. Mean Number of Errors to Criterion as a Function of Drug Dosage and Hours of Deprivation

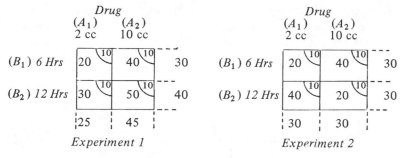

	Drug (A_1) 2 cc	Drug (A_2) 10 cc	
(B_1) *6 Hrs*	20	40	30
(B_2) *12 Hrs*	30	50	40
	25	45	

Experiment 1

	Drug (A_1) 2 cc	Drug (A_2) 10 cc	
(B_1) *6 Hrs*	20	40	30
(B_2) *12 Hrs*	40	20	30
	30	30	

Experiment 2

Both experiments investigate the effects of two levels of drug dosage (2 cc and 10 cc) and two levels of food deprivation (6 hours and 12 hours) on maze-learning performance in rats. The numbers in the cells represent mean number of errors to a specified criterion. The numbers in the corners of each cell represent the number of subjects in that cell. For Experiment 1, we can look first at the effects of drugs, averaged over the deprivation conditions. The average for all of the rats who were administered 2 cc of the drug is 25 errors. The average for all of the rats who had 10 cc of the drug is 45 errors. These means represent what is known as the main effect of drugs, because the effect of deprivation has been averaged out. Similarly, we can determine that the mean number of errors for all rats deprived for 6 hours is 30 errors and that for rats deprived 12 hours the corresponding value is 40 errors. These means represent the main effect of deprivation. There is no interaction because the higher level of deprivation produces more errors at both levels of drug dosage and vice versa. If these data are graphed (see Figure 9.1a), the two trend lines are parallel. (See *Notes on Interactions*, p. 290, for information on graphical methods.)

The situation is quite different in our hypothetical Experiment 2. In this experiment, rats given either 2 cc or 10 cc of the drug averaged 30 errors. Similarly, each of the deprivation groups averaged 30 errors. Neither drug dosage by itself nor deprivation level by itself had a differential effect on number of errors. Thus there is no significant main effect for either drug level or deprivation level. However, we do have an interaction. The rats deprived for 12 hours made more errors than the rats deprived for 6 hours when the drug dosage was 2 cc, but they made fewer errors than the rats deprived for 6 hours when the drug dosage was 10 cc. It should be noted that we might also have said that the rats receiving 2-cc dosages made fewer errors than the rats receiving 10 cc when the deprivation level was 6 hours, but that they made more errors when the deprivation level was 12 hours. Both statements are correct alternative ways of saying the same thing, and both indicate that the effect of one variable depends on the specific level of the other variable. Graphical presentation of these data (see Figure 9.1b, Experiment 2) shows that the trend lines are non-parallel.

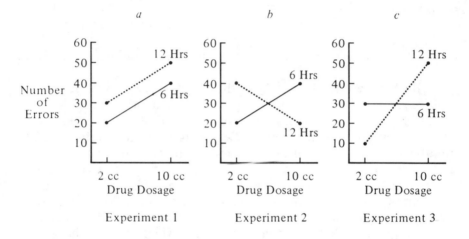

Figure 9.1. *Mean number of errors to criterion as a function of drug dosage and hours of deprivation.*

When any interaction is significant, interpretation is greatly aided by graphic presentation. In Experiment 1, with two significant main effects and no interaction, the experimenter may conclude that both increased deprivation and increased drug dosages, within the values studied, caused performance to deteriorate. In Experiment 2, with a significant interaction, you could conclude that neither drugs nor deprivation alone had a simple effect on performance but that the effect depended on the combination of the two variables. The larger drug dosage worsened performance under the lower level of deprivation but improved it under the higher level of deprivation.

A more difficult—and more common—situation is illustrated in Figure 9.1c for hypothetical Experiment 3, in which both the effects of drug dosage and the interaction of drug and deprivation are significant. In this case, the experimenter would be remiss if he concluded, on the basis of a significant main effect for drugs, that there was a general tendency for higher drug levels to produce more errors. Inspection of Figure 9.1c indicates that this effect occurs only after 12 hours of deprivation. In other words, the significant interaction may prevent improper interpretations of main effects.

A reasonable interpretation of Experiment 3 might be: "for the rats deprived 6 hours, drug dosage did not significantly affect performance. However, rats deprived 12 hours made more errors with 10 cc of the drug than with 2 cc." (A glance ahead at Figure 9.4c may suggest that some interpretations, such as this one, are more readily suggested by one configuration of the data than by another.)

If you wish to determine whether the difference between the 12-hour/10 cc condition and the 12-hour/2 cc condition was statistically significant, Tukey's (a) test may be performed. Experiment 3 demonstrates the gains in efficiency and knowledge possible with a factorial experiment. If you had investigated only 6-hour-deprived rats, you would have found no difference and might have reasonably concluded that the drug was ineffective. Had you investigated only 12-hour-deprived rats, you would have concluded that the drug was detrimental and might have concluded that this was a general effect. By looking at both of these independent variables simultaneously, you learned that the drug may be detrimental, but only under certain circumstances.

a X *b* *(between-subjects) ANOVA*. The *a* X *b* ANOVA is the general form of the two-way analysis of variance. This between-subjects analysis has *a* levels of one independent variable, where *a* is two or more, and *b* levels of a second independent variable, where *b* is two or more. Since the main effects are interpreted in exactly the same way as in the one-way design (see p. 276), those examples are not repeated here.

The interactions are generally interpreted in the same way as was indicated for the 2 X 2 ANOVA, and this section (see p. 280) should be reviewed for examples and general considerations. As either *a* or *b* or both become large, interpretations become more complex and difficult. The straightforward visual interpretations of interactions possible with a 2 X 2 design become difficult, if not impossible, as the number of levels of one or both variables increases. Nevertheless, significant interactions should generally be graphed as an aid to interpretation.

Increasing the number of levels beyond two for either independent variable has both advantages and disadvantages for the researcher. On the positive side, it becomes possible for him to find curvilinear relationships existing among his variables. In a curvilinear relationship, high and low values along the independent variable have a similar effect on the behavior, whereas intermediate values have a different effect. Such relationships cannot be found in cases of two levels of the independent variable, since all effects are linear when there are two levels of the independent variable.

On the negative side, the interpretation of interactions becomes increasingly difficult as more levels of the independent variables are added to the research design. The additional levels make interpretations of main effects more difficult as well, but they make interpretation of significant interactions particularly bothersome. Figure 9.2 shows the results of a hypothetical study that, again, combines levels of deprivation and drug dosage in a factorial design. In this case, however, there are four levels of drug dosage and five levels of deprivation.

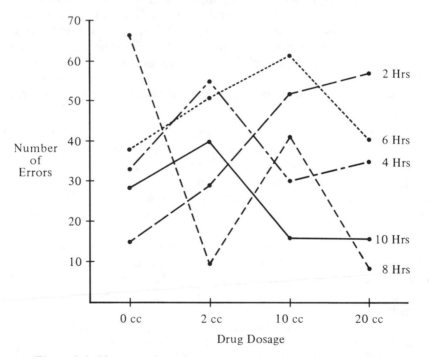

Figure 9.2. *Mean number of errors to criterion as a function of drug dosage and hours of deprivation.*

The data shown in Figure 9.2 are a nightmare. The poor experimenter almost certainly would be at loss to explain the meaning of a significant interaction. You may certainly say that the effect of drug dosage depends on the specific level of deprivation, but you would find it difficult to formulate a general principle describing a systematic effect. Such results are most likely to occur when little is known about the effects of the independent variable on behavior and when little is known about appropriate controls. Experiments should probably be kept simple, especially when research is in an exploratory stage.

a X *b* X *s (within-subjects) ANOVA*. Just as the *a* X *s* ANOVA is for practical purposes a one-way analysis, the *a* X *b* X *s* is for practical purposes a two-way analysis.[7] Interpretational problems are similar to those encountered in between-subjects two-way analyses (*a* X *b* ANOVA). If both independent variables have only two levels, visual interpretations are possible. As the number of levels increases, however, specific-comparison tests may be necessary to determine which effects are significantly different. If an interaction is significant, graphing the interaction should be undertaken before beginning any interpretation of the meaning of the results. Again, multilevel experiments may be difficult to interpret, particularly if the interaction of the two independent variables is significant.

a X *b mixed designs*. Although mixed designs are conceptually more complex than pure between-subjects, within-subjects, or simple factorial designs, the researcher encounters no special interpretational problems with them. The same basic principles presented for the between- and within-subjects designs may be used in interpreting these more complex designs. Remember that the power of the within-subjects portion of the design is always greater than the power of the between-subjects portion of the design. This means that the possibility of detecting real differences when they exist is always greater in the within-subjects portion of this design. Therefore it is always more likely that differences will be found in the within-subjects than in the between-subjects portion of the analysis, even when equally large differences exist in the population. If yours is a 2 X 2 design, review the section on 2 X 2 ANOVA. If your design has multiple levels, also read the section on *a* X *b* ANOVA.

[7]The analysis again partials out the variance due to subjects and tests the effects of the independent variable by using an interaction of treatments with subjects as the error term. See Footnote 6, p. 278.

C. ANALYSES WITH THREE INDEPENDENT VARIABLES

Between-Subjects Designs: 2 × 2 × 2, *a* × *b* × *c*
Within-Subjects Designs: a × *b* × *c* × *s*
Mixed Designs: a × *b* × *c (B-B-W, B-W-W)*
Interpretational Steps
IF THE *F* RATIO FOR *A* × *B* × *C* INTERACTION *IS NOT* SIGNIFICANT:
follow through the steps spelled out for *Analyses with Two Independent Variables*, p. 279.
IF THE *F* RATIO FOR *A* × *B* × *C* INTERACTION *IS* SIGNIFICANT:
1. Compute means for each cell of the *A* × *B* × *C* table.
2. Graph the means (see *Note on Interactions*, p. 290) and visually inspect the meaning of the interaction. If visual inspection does not clarify the meaning of the interaction, Tukey's (a) test may be performed to identify the locus of the significant difference (see Chapter 10, p. 319).
3. Follow through the steps spelled out for *Analyses with Two Independent Variables*, p. 279, *except:*
4. Interpret each of your significant two-way interactions and each of your significant main effects with caution; indicate the limitation on the generality of your statement introduced by the significant triple interaction. Remember that the significant triple interaction almost always limits the generality of statements about both the two-way interactions and the main effects (see *Note on Interactions*, p. 290).

GENERAL CONSIDERATIONS AND EXAMPLES

The 2 × 2 × 2 ANOVA (between-subjects). The 2 × 2 × 2 design is the simplest experimental design in which three independent variables may be tested simultaneously. The 2 × 2 × 2 ANOVA is a between-subjects design with two levels of each of three factorially combined independent variables. The analysis allows us to test for the effects of (1) all three input dimensions separately (main effects), (2) three different two-way interactions, and (3) the three-way interaction.

As indicated in the *Interpretational Steps*, the interpretation proceeds backwards, from the more complex effects to the simpler. If the three-way interaction is significant, it should be graphed as two two-way interactions (see *How to Plot the Data*, p. 294). The three-way interaction can be presented graphically in six different ways. See *Note on Interactions*, p. 293, for details on how the three-way interactions may be plotted. All six methods display the same information somewhat differently, but sometimes one display will be confusing, whereas a second display shows the meaning of the inter-action clearly. If your first graphic presentation does not lead to a clear interpretation, try a second, and so on, until you find the one that makes clear sense to you.

A significant three-way interaction indicates that the two-way interactions are different for the two levels of the third independent variable. Figure 9.3 illustrates a hypothetical three-way interaction. It may be seen from Figure 9.3 that there is no interaction (shown by parallel lines) between drug dosage and hours of deprivation when the study is performed with a 10-db background noise. However, when

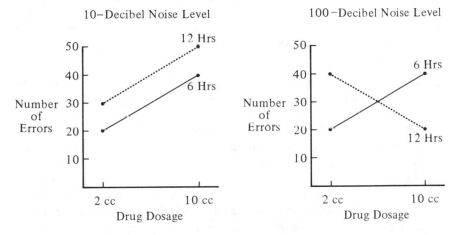

Figure 9.3. *Mean number of errors to criterion as a function of drug dosage, hours of deprivation, and noise level.*

the noise level is raised to 100 db, drug dosage and level of deprivation show a strong interaction. If the three-way interaction is significant, the two-way interactions and the main effects must be interpreted with the meaning of the three-way interaction in mind. The main effects may have to be further qualified if the two-way interactions are also significant. Tukey's (a) test for interaction may be needed to determine which specific levels in the interaction are producing the significant difference.

a X *b* X *c (between-subjects) ANOVA*. The *a* X *b* X *c* design is the general form of the three-way design. As the number of levels of any or all of the three independent variables becomes greater than two, the interpretational problems increase, just as they did in the *a* X *b* design. With multiple levels of the three independent variables, the possibility of uninterpretable interactions increases. Not only must you worry about uninterpretable two-way interactions, but the three-way interaction may also be uninterpretable. In a three-way or higher-order design, it is particularly important that only those levels of each independent variable indicated by practical or theoretical considerations be included. Again, you should graph all interactions and use specific-comparison tests when you need them.

a X *b* X *c* X *s (within-subjects) ANOVA*. For within-subjects designs with three factorially combined independent variables, the *a* X *b* X *c* X *s* analysis of variance is appropriate. Again, this is a four-way design that partials out the variance due to subjects but does not test it; therefore it may be treated as a three-way design. Interpretational considerations are identical to those for the 2 X 2 X 2 and the *a* X *b* X *c* ANOVAs.

a X *b* X *c mixed designs*. Although mixed designs are conceptually more complex than pure between-subjects, within-subjects, or simple factorial designs, the same principles presented for the between- and within-subjects designs generally may be used in interpreting these more complex designs. Remember, of course, that the power of the within-subjects portion of the design is always greater than the power of the between-subjects portion of the design. Therefore it is always more likely that differences will be found in the within-subjects portion than in the between-subjects portion of the analysis, even when equally large differences exist in the population.

D. ANALYSES WITH FOUR INDEPENDENT VARIABLES

(Higher-order designs proceed analogously.)
Interpretational Steps
IF THE *F* RATIO FOR THE *A* X *B* X *C* X *D* INTERACTION *IS NOT* SIGNIFICANT OR HAS BEEN POOLED:
1. Follow through the steps spelled out for *Analyses with Three Independent Variables*, p. 286.
IF THE *F* RATIO FOR THE *A* X *B* X *C* X *D* INTERACTION *IS* SIGNIFICANT:
1. Compute the means for each cell of the *A* X *B* X *C* X *D* table.
2. Graph the means (see *Note on Interactions*, p. 290) and visually inspect the meaning of the interaction. If visual inspection does

not clarify the meaning of the interaction, Tukey's (a) test may be performed to identify the locus of the significant differences (see Chapter 10, p. 319).

3. Follow through the steps spelled out for *Analyses with Three Independent Variables*, p. 286, *except:*

4. Interpret each of your significant two- and three-way interactions and each of your significant main effects with caution, qualifying them by giving the limitation on your statement introduced by the significant quadruple interaction. Remember that the significant quadruple interaction (if interpretable at all) limits the generality of statements about the lower-order interactions and the main effects.

GENERAL CONSIDERATIONS AND EXAMPLES

Factorial-design experiments can, of course, be expanded endlessly. Theoretically, there is no limit to the number of independent variables that may be studied simultaneously or to the number of levels for each independent variable. No matter how many independent variables there are or how many levels of each, analysis of variance can be expanded to analyze the results, using the same simple computing rules and formulas appropriate for simple designs. We have pointed out, however, that practical considerations set limitations (if theoretical considerations do not) on the complexity of factorial designs. The more independent variables and the more levels of independent variables that are studied, the more complex the interpretation becomes.

A FINAL NOTE ON INTERPRETATION

When you design a study, your major concern is to be able to provide an unequivocal answer to your research question. You should, in general, construct the simplest study that will answer the question(s) you wish to ask. Be sure that you can justify every independent variable and every level of each independent variable. Your ability to perform a complex analysis is not an adequate rationale for performing a large study.

Many problems do demand complex research designs (and, for these, complex research designs should be employed), but inexperienced researchers too often believe that the "better experiment" is the one with more factors and more levels of each factor. It is foolish to select six levels of an independent variable when two would answer your question. Not only may you "swamp" an interesting difference (the difference may not show up because there are too many levels of the independent variable); you also create problems of interpretation not present in a simpler experiment. Interpretation of research results almost always becomes more complex as the design of the research becomes more complex. Remember that brevity is the soul of wit, and simplicity is often the key to elegant and easily interpreted research.

NOTE ON INTERACTIONS

Figure 9.4 summarizes a variety of two-way interactions. In each case, there are two levels of each factor. When there are two levels of

Figure 9.4. *Typical interactions.*

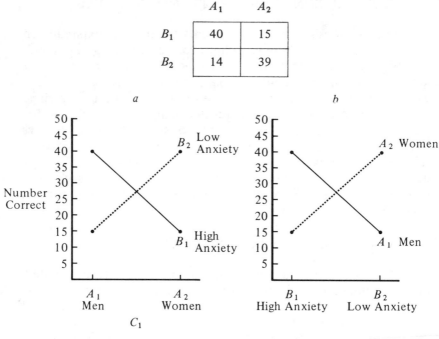

Significant Interaction
No Significant Main Effects

Figure 9.4 (continued). *Typical interactions.*

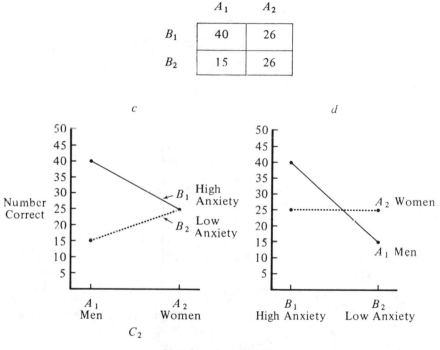

	A_1	A_2
B_1	40	26
B_2	15	26

C_2

Significant Interaction
Significant Main Effect of B

Figure 9.4 (continued). *Typical interactions.*

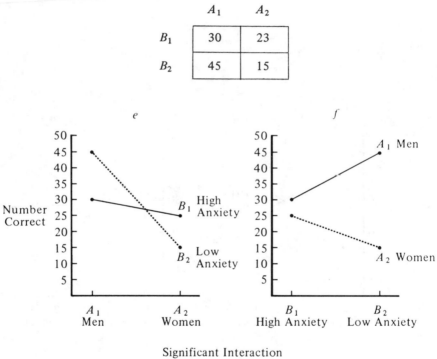

Significant Interaction
Significant Main Effect of *A*

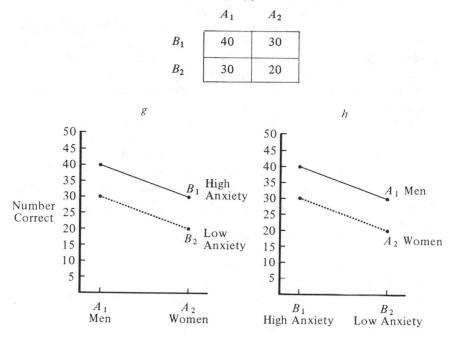

Figure 9.4 (continued). *Typical interactions.*

No Interaction
Both Main Effects Significant

one factor and three or more of the second, graphical representation is still relatively straightforward. With three or more levels of both factors, representative graphical presentation becomes impractical, and we have left it to you to generalize these cases from the two-way presentation. A number of major principles may be illustrated with this set of figures.

HOW TO PLOT THE DATA

1. Begin by drawing your horizontal and vertical axes. Label the vertical axis with the dependent variable. Select one of the independent variables and label two (or more, if necessary) hatch marks on the horizontal axis with these labels. (Suppose you label the axis with A_1 and A_2.) Then go to your data table and select *one level of B.* Plot the data points for A_1B_1 and A_2B_1. Connect these two points with a line and label this B_1. It is essential that

you connect these lines correctly. You cannot interpret your
results properly if you do not. Repeat the same procedure for B_2.
You now have two lines, one representing B_1 and one representing
B_2.

2. Note (Figure 9.4c and d, e and f, g and h) that all data (involving
two factors) can be presented in two alternative ways. Until you
begin to understand the relationship between the patterns, you
will do well to plot the data both ways.

3. *Three-way designs.* Plotting a three-way interaction involves plotting
the two-way interaction at each level of the third variable. For
example, look at Figure 9.4a and c. We have indicated that 9.4a
shows the data for level C_1 and 9.4c shows the data for level C_2.
Since the two curves are not similar, we would expect a significant
triple interaction.

HOW TO INTERPRET THE INTERACTIONS

The interpretation of interactions involves not just a statement
of the result but also the qualification, if necessary, of the main
effects as a result of the significant interaction. A caution: remember
that your significant interaction indicates only that the effect of one
variable depends on the level of the other variable. The interaction
does *not* mean, by itself, that any of the specific differences within
the interaction table are significant; your interpretations should
avoid making this implication. If you wish to say that one treatment
combination differs from another treatment combination, you must
perform a specific-comparison test. (In all interpretations below we
assume that specific-comparison tests have been performed.) The
following examples provide interpretations for each of the graphs in
Figure 9.4. In each case, we have provided a general interpretation
and then an interpretation that assumes that Factor A is gender:
men $= A_1$, women $= A_2$; Factor B is anxiety level: high anxiety
$= B_1$, low anxiety $= B_2$; Factor C (Figures 9.4a and c only) is task
difficulty: easy task $= C_1$, difficult task $= C_2$. The dependent
variable is number of items correct. We hope it is evident that the
factors are randomly assigned to a variety of curves and that none
of the interpretations is psychologically meaningful except by
accident.

1. *Figure 9.4a and b. Significant interaction, no main effect.* The
significant interaction indicates that Condition B_1 is superior to
Condition B_2 for Condition A_1, but that Condition B_2 is superior
to Condition B_1 for Condition A_2. There are no significant main
effects.

Specific example. High-anxious men perform better than low-anxious men, but low-anxious women perform better than high-anxious women. OR: High-anxious men perform better than high-anxious women, but low-anxious women perform better than low-anxious men.

2. *Figure 9.4c and d. Significant interaction, significant main effect of B.* From c: although B_1 is superior to B_2, this superiority is due solely to differences at A_1, since the two levels of B do not differ at A_2. From d: A_1 is superior to A_2 for Condition B_1, but A_2 is superior to A_1 for Condition B_2. (Note that it is not as easy to see in the latter array that the superiority of B_1 is due solely to differences in A_1.)

Specific example. From c: although high-anxiety subjects perform better than low-anxiety subjects, this superiority is due solely to differences among men, since women of both anxiety levels performed the same. From d: high-anxiety men perform better than high-anxiety women, but low-anxiety women perform better than low-anxiety men. (Note, again, that in this latter array it is not as easy to see that the main effect for anxiety is due only to differences in men.)

3. *Figure 9.4e and f. Significant interaction, significant main effect of A.* From e: although Condition B_2 is superior to Condition B_1 for Condition A_1, Condition B_1 is superior to Condition B_2 for Condition A_2; however, Condition A_1 is always superior to A_2. From f: Condition A_1 is superior at all levels to Condition A_2, but these differences are greater at Condition B_2 than at Condition B_1.

Specific example. From e: men perform better than women at both levels of anxiety; however, low-anxiety men perform better than high-anxiety men, and high-anxiety women perform better than low-anxiety women. From f: men perform better than women at both levels of anxiety; however, the difference is greater for the low-anxiety condition than for the high-anxiety condition.

4. *Figure 9.4g and h. No interaction; significant main effects of both A and B.* Condition B_1 is superior to Condition B_2 and Condition A_1 is superior to Condition A_2. There was no interaction.

Specific example. High-anxiety subjects perform better than low-anxiety subjects and men perform better than women. There was no interaction between the variables.

5. *Figure 9.4a and c. Significant triple interaction.* The significant three-way interaction indicates that the relationship between A

and B differs for the two levels of C. For Condition C_1, Condition B_2 is superior to Condition B_1 at A_1, but Condition B_1 is superior to Condition B_2 at A_2. However, for Condition C_2, B_1 is superior to B_2 at A_1, but the two levels of B do not differ at A_2.

Specific example. The significant three-way interaction indicates that the relationship between gender and anxiety differs for the easy task and the difficult task. On the easy task, low-anxious men perform better than high-anxious men, but high-anxious women perform better than low-anxious women. On the difficult task, high-anxious men perform better than low-anxious men, but women of both anxiety levels perform the same.

CHAPTER TEN

SPECIFIC COMPARISONS[1]

An overall test of significance, like those presented in the preceding chapters, tells you whether there are any overall differences *among* the groups or conditions studied. The null hypothesis tested is that *all* of the groups or sets of measurements are drawn from the same population. The overall test of significance does not provide information about differences between specific groups or sets of measurements in your experiment—*except* when there are only two levels of an independent variable. When there are two levels, the overall test of significance and a specific comparison are identical, and these exhaust the possible comparisons. As the number of levels increases, and particularly when more than one independent variable has been included in the design, the number of possible comparisons between and among specific conditions increases. Most often, the comparisons of interest are between groups or conditions taken two at a time. However, a variety of more complex comparisons are possible, particularly for score data. Most complex comparisons lie beyond the scope of a brief statistics manual. A variety of the simpler ones are included here, however, since the overall test often does not permit you to make reasonable interpretations of your data or to draw conclusions about your hypothesis when you have a specific experimental prediction.

There has been extensive debate about the proper method for making specific comparisons within an experiment or study. An understanding of these issues may help you decide which methods to apply.

Error rate. Much of the lengthy and confusing statistical debate[2] about appropriate specific-comparison techniques concerns the problem of controlling the α rate when comparisons between different groups or conditions are made. If there are only two conditions, there is only one possible comparison that can be made; and, as we have indicated before, the overall statistic indicates whether the two groups

---

[1] Another name for these methods is multiple comparisons.

[2] A brief, clear summary of many of the relevant issues is given by Ryan (1959) and reprinted in Kirk (1972). The review by Games (1971) is also excellent.

are significantly different. As the number of conditions increases, the number of possible comparisons increases very rapidly. When you have a treatments, the formula $a(a - 1)/2$ gives the total number of different comparisons possible for treatments taken two at a time. Whereas the number of comparisons when there are only two levels is one, it is ten if there are five levels and 45 is there are ten levels.

As the number of possible comparisons increases, the probability of finding differences significant by chance also increases. If we do an experiment in which there is only one comparison, our alpha level accurately reflects the probability of making a Type I error. If two comparisons are involved, the situation changes somewhat. There are now four possible outcomes. We can make a Type I error in both comparisons, in the first comparison but not in the second, in the second but not in the first, or in neither. If we assume that the comparisons are *independent* and set our alpha level at .05, we can determine the probability that each of these outcomes will occur. The probability of a Type I error in both comparisons is .05 $\times$ .05 = .0025. The probability of a Type I error in the first comparison but not in the second is .05 $\times$ .95 = .0475. Similarly, the probability of a Type I error in the second but not in the first is .95 $\times$ .05 = .0475. If we add these three figures together, we obtain the probability of committing at least one Type I error in our experiment. The combined probability is .0975.

A simpler way of computing the probability involves multiplying the probability of *not* making an error in the first comparison (.95) by the probability of *not* making one in the second (.95), and subtracting the product from 1 (.95 $\times$ .95 = .9025; 1 $-$ .9025 = .0975). Similarly, the probability of making *at least one Type I error* in a series of *independent* comparisons can be determined by taking $(1 - \alpha)$, raising it to the power of the number of comparisons, and subtracting this value from 1. For example, for $\alpha = .05$, the probability of making at least one Type I error if there are ten independent comparisons is $1 - .95^{10} = 1 - .5987 = .4013$. If the comparisons are not independent, as often happens, the situation becomes more complex. Depending on the type of nonindependence, the probability of at least one Type I error can be either greater or less than the probability of an error if the comparisons are independent. However, it is clear that, when many comparisons are involved, the probability of committing a Type I error in the *experiment* is considerably greater than the alpha level for each individual comparison.

The problem that has plagued statisticians should be apparent at this point. If the *error rate per comparison* is held constant, for example, at .05, the probability of making a Type I error somewhere in the experiment increases as the number of comparisons increases. On the other hand, if the *experimentwise error rate*[3] is held constant at .05 (for example, by making your comparisons at α levels higher than .05), the probability of making Type II errors greatly increases. Persuasive arguments can be found for practically every conceivable strategy—from doing the most powerful, least conservative tests that hold error rate constant per comparison (*t* tests for all pairs of means, or a 2 X 2 χ^2 to partition a large contingency table) to doing the least powerful, most conservative tests that hold experimentwise error rate constant (Scheffé's test, Ryan's procedure). Many statisticians have offered compromise procedures that fall somewhere between these two extremes, and these tests have become very popular.

It is extremely difficult to make recommendations about the error rate that you should adopt for your study. More than anything, deciding on error rate is a matter of establishing conventions, and no clear conventions have been established for specific comparisons. Since the majority of statisticians do seem to agree that controlling error rate per comparison is too liberal a criterion, we have presented relatively conservative procedures (high protection against Type I errors). We also refer to alternative procedures.

A priori (planned) vs a posteriori (post mortem) comparisons. Statistical literature has generally distinguished between cases in which the researcher has developed an *a priori* hypothesis for predicting a difference between any two groups and those in which he has developed the hypothesis as a result of looking at his data (*a posteriori* or *post mortem*). In general, when *a priori* hypotheses have been developed, it has been concluded that pairwise comparisons may be made by the use of simple procedures that do not attempt to adjust the experimentwise error rate (*t* tests, rank sums, 2 X 2 χ^2, and so on). For *a posteriori* tests, the type of procedures that we have included in this chapter are generally recognized to be required. We recommend that beginners use the more conservative procedures presented in this manual for all comparisons until they have had an opportunity to read the complex literature on specific comparisons, understand the issues, and can make informed decisions of their own.[4]

This chapter presents a series of closely related tests. Ryan's procedure (frequency and ordered data) is, in fact, a variation on Tukey's (a) test (score data). Newman-Keuls' (score data) test bears a family relationship to Tukey's. Nemenyi's (ordered data) test,

---

[3]The experimentwise error rate is the probability that one or more Type I errors will be made in an experiment.
[4]For diverse points of view on the issue, see Kirk (1972), Chapter 8.

however, is only distantly related to the others. Once the general
procedure for one of these is understood, the others will be relatively
easy to perform.

A. FREQUENCY DATA

The simplest and most straightforward technique for doing
specific comparisons within a χ^2 table is Ryan's procedure (see Ryan,
1960). Ryan's procedure allows for all pairwise comparisons between
proportions within an $a \times 2$ table by adjusting the significance levels
for specific comparisons *to keep experimentwise error rate constant.*
    The method is quite simple when one independent variable has
two levels (we discuss the other case in a later paragraph). For example,
in a 5×2 table with five levels of A and two levels of B, ten pairwise
comparisons may be made (see Table 10.1). In other words, there are
ten 2×2 χ^2s that may be performed. You determine the proportion
of each of the five levels of A that are B_1 (or B_2). The five levels of A
are then ordered according to the size of their proportions. In this

Table 10.1.

	A_1	A_2	A_3	A_4	A_5	
B_1	6	3	5	4	8	26
B_2	4	7	5	6	2	24
	10	10	10	10	10	50

case, the ordering would be A_5, A_1, A_3, A_4, and A_2. $2 \times 2 \chi^2$s are
then computed on the series of comparisons and tested against an
adjusted significance level.
    Ryan's procedure thus involves computing a series of already
familiar 2×2 χ^2s and testing them against a tabled χ^2 whose value
is determined by α, a, and d, where α is the level we wish for the
experiment as a whole, a is the number of levels of the dimension
with more than two levels, and d is the number of ordered levels
spanned by the comparison (in this example, the comparison between
A_5 and A_2 spans five levels and $d = 5$; the comparison between A_1
and A_4 spans three levels and $d = 3$; and so on. If your value exceeds
the tabled value, the difference is significant at an experimentwise
.05 level.
    Ryan's procedure cannot be easily applied to χ^2 tables that are
larger than $a \times 2$. One solution is to either drop out or combine

categories until an $a \times 2$ table is obtained and then perform Ryan's procedure. A second solution is to use some other specific-comparison technique, such as orthogonal comparisons (see Castellan, 1965, and Bresnahan & Shapiro, 1966).

Description:

Allows pairwise comparisons in χ^2

| RYAN'S PROCEDURE |
| FREQUENCY DATA |

Two levels of one independent variable AND any number of levels of the second independent variable

α is controlled experimentwise

Overall χ^2 for the $2 \times a$ table is significant

Follow through the steps below.

- *Analysis is appropriate when the overall χ^2 for the $2 \times a$ contingency table is significant.*

- *If the contingency table is larger than $2 \times a$, see p. 300.*

- *There are a possible $a(a - 1)/2$ pairwise comparisons that may be made.*

A. DESCRIPTION OF THE STUDY

A random sample of freshmen, sophomores, juniors, and seniors were asked whether they agreed with or disagreed with a proposed change in student government. The frequency of agreement or disagreement by class standing was tested by an overall χ^2. The overall χ^2 for this contingency table, 14.30, was significant.

Therefore the null hypothesis that there was no association between class standing and agreement on this issue was rejected. The student council wished to know how the specific classes differed in their support of the changes.

B. RECORDING THE DATA

1. The data should be recorded in the same $2 \times a$ contingency table used for computing the overall χ^2 (see Table 10.2). A is always the independent variable with $a > 2$ levels, and B the independent variable with 2 levels.

Table 10.2

	A_1 Freshman	A_2 Sophomore	A_3 Junior	A_4 Senior	
Agree (B_1)	5	13	8	6	32
Disagree (B_2)	16	3	5	9	33
	21	16	13	15	65

C. COMPUTATIONS

1. From Appendix 5, for your value of a, find the tabled values of χ^2 for each value of $d - 1$, where d is the number of ordered levels spanned by the comparison. (Note that d changes for each comparison.) For convenience, list the values of χ^2 in descending order.

> For $\alpha = .05$ and $a = 4$, tabled values of χ^2:
> for $d - 1 = 3$, $\chi^2 = 6.97$; for $d - 1 = 2$, $\chi^2 = 6.25$;
> for $d - 1 = 1$, $\chi^2 = 5.02$

2. Compute the proportion of B_1 (or B_2) at each level of A; that is, compute $A_1 B_1 / A_1$, $A_2 B_1 / A_2$, $A_3 B_1 / A_3$, and so on, from the original data.

> Proportion of freshmen who agree $= A_1 B_1 / A_1 = 5/21 = .238$
> Proportion of sophomores who agree $= A_2 B_1 / A_2 = 13/16 = .812$
> Proportion of juniors who agree $= A_3 B_1 / A_3 = 8/13 = .615$
> Proportion of seniors who agree $= A_4 B_1 / A_4 = 6/15 = .400$

3. Order the proportions from large to small.
   a. If you have four or more levels of the independent variable, it may be convenient to summarize the comparisons in a table. Prepare the table with the proportions arranged in ascending order along the top (column heads) and along the sides (row heads) of the table. In the body of the table, list the χ^2s as they are computed (Step #4). See Table 10.3.

Table 10.3

	Sophomores .812	Juniors .615	Seniors .400	Freshmen .238	d	d − 1	χ^2_{Tabled}
Sophomores .812		—	3.95	9.80			
					4	3	6.97
Juniors .615			—	3.37			
					3	2	6.25
Seniors .400				—			
					2	1	5.02
Freshmen .238							

b. If you have only three levels of a, the table may not be useful.
4. A 2 × 2 χ^2 may be performed to compare each pair of proportions indicated in the table. (See Chapter 5, p. 69, for computing rules for 2 × 2 χ^2.) Placing the tests in a prescribed order prevents unnecessary computations and prevents inconsistencies in interpretation.
   a. The first 2 × 2 χ^2 must be performed on the most extreme proportions. Enter the value of χ^2 in the table at the appropriate intersection at the upper right of the first row of the table (see Table 10.3). *If this χ^2 is significant*, move one intersection to the left on the first row and perform the χ^2 appropriate for this intersection. Continue performing χ^2s, moving one intersection to the left on the first row, until you encounter a nonsignificant χ^2. Then begin on the right of the second row and begin performing χ^2s in that row until you encounter a χ^2 that is not significant *or* you come to a column in which there was not a significant difference in the first row. Then begin at the right of the third row, and so on. If the χ^2 at the right of any row is not significant, the testing ends, for

there can be no further significant differences between proportions in either rows or columns of the table.

b. Every χ^2_{Obtained} is tested against χ^2_{Tabled} (from #1) for a and $(d - 1)$, where d is the number of ordered levels spanned by the comparison (Appendix 5). Table 10.3 indicates which χ^2_{Tabled} corresponds to each χ^2_{Obtained}.

c. If χ^2_{Obtained} exceeds the appropriate χ^2_{Tabled}, there is a significant difference between proportions.

All these comparisons are based on values in Table 10.2, and the resulting χ^2s are entered in Table 10.3.

Row 1: Most extreme proportion
Sophomores vs freshmen:

	Sophomores	Freshmen	
Agree	13	5	18
Disagree	3	16	19
	16	21	37

$$\chi^2 = \frac{n(|bc - ad| - n/2)^2}{(a + b)(c + d)(a + c)(b + d)} = \frac{37(|15 - 208| - 37/2)^2}{(18)(19)(16)(21)}$$

$$= \frac{1,126,659.25}{114,912} = 9.80$$

$\chi^2_{\text{Tabled}} = 6.97$, for $\alpha = .05$, $a = 4$, and $d - 1 = 3$. $9.80 > 6.97$. Therefore the difference between the proportions is significant. (Perform another χ^2 on the same row.)

Row 1: Second comparison
Sophomores vs seniors:

	Sophomores	Seniors	
Agree	13	6	19
Disagree	3	9	12
	16	15	31

$$\chi^2 = \frac{31(|18 - 117| - 31/2)^2}{(19)(12)(16)(15)} = \frac{216,139.75}{54,720} = 3.95$$

$\chi^2_{\text{Tabled}} = 6.25$, for $\alpha = .05$, $a = 4$, and $d - 1 = 2$. $3.95 < 6.25$. Therefore the difference between proportions is not significant. (Begin on next row.)

Row 2: First comparison
Freshmen vs juniors:

	Freshmen	Juniors	
Agree	5	8	13
Disagree	16	5	21
	21	13	34

$$\chi^2 = \frac{34(|128 - 25| - 34/2)^2}{(13)(21)(21)(13)} = \frac{251,464}{74,529} = 3.37$$

$\chi^2_{\text{Tabled}} = 6.25$, for $\alpha = .05$, $a = 4$, and $d - 1 = 2$. $3.37 < 6.25$. Therefore the difference between the proportions is not significant. (End computations.)

D. INTERPRETATION

There is an overall association between class standing and opinion on the changes in student government. However, the only two classes that differ significantly are freshmen and sophomores, with freshmen more likely to disagree with the changes and sophomores more likely to agree.

B. ORDERED DATA

Ryan's procedure (described in the section on frequency data) is also appropriate for pairwise comparisons when the Kruskal-Wallis test has been used to analyze ordered data. The procedure for applying this test is virtually identical to the procedure used for frequency data.

To use Ryan's procedure, the medians of the groups or sets of measurements are ordered. Ryan's procedure involves computing a series of already familiar rank-sums tests and testing them against a tabled value of Z, whose value is determined by α, a, and d, where α is the level we wish for the experiment as a whole, a is the number of levels of the independent variable, and d is the number of ordered levels spanned by the comparison. The precise meaning of d is explained in the examples. If your value exceeds the tabled value, the difference is significant at an experimentwise .05 level.

Ryan's procedure is also appropriate for the Friedman test, but the extended tables necessary for evaluating it have not been developed; therefore we will present an alternative test, Nemenyi's test, to permit specific comparisons for the Friedman test. Nemenyi's test controls experimentwise error rate but involves somewhat different assumptions than Ryan's procedure.

RYAN'S PROCEDURE **ORDERED DATA** **(KRUSKAL-WALLIS)**

Description:

Allows pairwise comparisons in Kruskal-Wallis
a levels of the independent variable
α is controlled experimentwise
Overall H is significant
For specific comparisons following the Friedman test, see Nemenyi's test, p. 311

Follow through the steps below.

- *Analysis is appropriate when the overall H is significant.*

- *There are a possible a (a − 1)/2 pairwise comparisons that may be made.*

A. DESCRIPTION OF THE STUDY

A marine biologist was interested in determining whether temperature affected the rate at which blue-belly snapper eggs hatch. He randomly selected 40 fertilized snapper eggs from the same female and placed ten of them in an aquarium heated to 90°, ten in an aquarium heated to 85°, ten in an aquarium heated to 80°, and ten in an aquarium heated to 75°. He assigned a rank of 1 to the first egg that hatched, 2 to the next egg, and 40 to the last egg. A Kruskal-Wallis test applied to the data yielded an H cf 17.40. With 3 df, an H of 7.815 is significant at the .05 level. He therefore rejected the null hypothesis that the four groups came from the same population and concluded that there was an overall relationship between water temperature and hatching rate. He wished to know which specific differences between temperatures were significant.

B. RECORDING THE DATA

1. The data should be recorded in the same table used to compute the overall H (see Table 10.4).

Table 10.4

	$90°$	$85°$	$80°$	$75°$
	1	2	4	13
	3	8	17	21
	5	9	19	23
	6	12	22	25
	7	18	26	27
	10	20	29	32
	11	24	30	35
	14	28	34	38
	15	31	36	39
	16	33	37	40
Total	88	185	254	293
Median	9	19	27.5	29.5

C. COMPUTATIONS

1. From Appendix 6, for your value of a, find tabled values of Z for each value of $d - 1$, where d is the number of ordered levels spanned by the comparison. (Note that d changes for each comparison.) For convenience, list the values of Z in descending order.

For $\alpha = .05$ and $a = 4$, tabled values of Z:

for $d - 1 = 3, Z = 2.64$; for $d - 1 = 2, Z = 2.50$;

for $d - 1 = 1, Z = 2.24$

2. Compute the median for each group and order the groups by increasing size.

   a. If you have four or more levels of the independent variable, you may find it convenient to summarize the comparisons in a table (see Table 10.5). Prepare the table with the medians arranged in ascending order along the top (column heads) *and* along the sides (row heads) of the table. In the body of the table, list the Zs as they are computed (Step #3).

Table 10.5

	$90°$ 9	$85°$ 19	$80°$ 27.5	$75°$ 29.5	d	$d - 1$	Z_{Tabled}
$90°$ (9)		-2.12	-3.17	-3.55			
					4	3	± 2.64
$85°$ (19)			$-$	-2.19			
					3	2	± 2.50
$80°$ (27.5)				$-$			
					2	1	± 2.24
$75°$ (29.5)							

   b. If you have only three levels of a, the table may not be useful.

3. A rank-sums test may be performed to compare every pair of medians indicated in the table. (See Chapter 6, p. 102, for computing rules for the rank-sums test). Doing the tests in the prescribed order prevents unnecessary computations and prevents inconsistencies in interpretations.

   NOTE: For each analysis, the ranks must be *reranked* from 1 to N (where $N = n_1 + n_2$) for the two groups on which the analysis is being performed.

   a. The first rank-sums test must be performed on the most extreme medians. Enter the value of Z in the table at the appropriate intersection at the upper right of the first row of the table (see Table 10.5). If this Z is significant, move one intersection to the left on the first row and perform the Z appropriate for that intersection. Continue performing Zs, moving one intersection to the left on the first row until you encounter a nonsignificant Z. Then begin on the right of the second row and begin performing Zs in that row until you encounter a Z that is not significant *or* you come to a column in which there was not a significant difference in the first row. Then begin at the right of the third row, and so on. If the Z at the right of any row is not significant, the testing ends, for there can be no further significant difference between medians in either rows or columns of the table.

b. Every Z is tested against the tabled Z (from Section 1) with the appropriate a and $(d - 1)$, where d is the number of ordered levels that is spanned by the comparison. Table 10.5 indicates which Z_{Tabled} corresponds to each $Z_{Obtained}$.

c. If $Z_{Obtained}$ exceeds the appropriate Z_{Tabled}, there is a significant difference between medians.

All these comparisons are based on values in **Table 10.4**, and the resulting Zs are entered in **Table 10.5**.

Row 1: Most extreme medians
90° vs 75°:

> NOTE: The ranks must be reranked from 1 to 20 for this comparison, since there are only 20 ranks involved in the comparison.

The 90° group has ranks of 1, 2, 3, 4, 5, 6, 7, 9, 10, 11.
The 75° group has ranks of 8, 12, 13, 14, 15, 16, 17, 18, 19, 20.
Total of ranks, 90° group = 58.
Total of ranks, 75° group = 152.
Rank-sums test:

$$Z = \frac{2T_i - n_i(N + 1)}{\sqrt{\dfrac{n_1 n_2 (N + 1)}{3}}} = \frac{2(58) - 10(21)}{\sqrt{\dfrac{(10)(10)(21)}{3}}} = -3.55$$

$Z_{Tabled} = 2.64$, for $\alpha = .05, a = 4, d - 1 = 3$.
3.55 > 2.64. Therefore there is a significant difference between medians for 90° and 75°. (Perform another rank-sums test on the same row.)

Row 1: Second comparison
90° vs 80°:

> NOTE: The ranks must be reranked from 1 to 20 for this comparison, since there are only 20 ranks involved in the comparison.

The 90° group has ranks of 1, 2, 4, 5, 6, 7, 8, 9, 10, 11.
The 80° group has ranks of 3, 12, 13, 14, 15, 16, 17, 18, 19, 20.
Total of ranks, 90° group = 63.
Total of ranks, 80° group = 147.
Rank-sums test:

$$Z = \frac{2(63) - 10(21)}{\sqrt{\dfrac{(10)(10)(21)}{3}}} = -3.17$$

$Z_{\text{Tabled}} = 2.50$, for $\alpha = .05, a = 4, d - 1 = 2$.
$3.17 > 2.50$. There is a significant difference between the medians for 90° and 80°. (Perform another rank-sums test on the same row.)

Row 1: Third comparison
90° vs 85°:
The rank-sums test is performed in the same fashion for this comparison. $Z = -2.12$.
$Z_{\text{Tabled}} = 2.24$, for $\alpha = .05, a = 4, d - 1 = 1$.
$2.12 < 2.24$. The difference between 90° and 85° is not significant. (Begin on next row.)

Row 2: First comparison
85° vs 75°:
The rank-sums test is performed in the same fashion for this comparison. $Z = -2.19$.
$Z_{\text{Tabled}} = 2.50$, for $\alpha = .05, a = 4$, and $d - 1 = 2$.
$2.19 < 2.50$. The difference between 85° and 75° is not significant. (End computations.)

D. INTERPRETATION

Overall, there is a significant relationship between water temperature and hatching rate of eggs. Eggs in 90° water hatched significantly more quickly than eggs in either 75° or 80° water. No other differences between groups were significant.

Description:

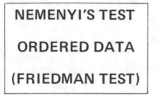

Allows pairwise comparisons
    following Friedman test
a levels of the independent
    variable
α is controlled experimentwise
Overall χ_r^2 is significant
For specific comparisons
    following Kruskal-Wallis test,
    see Ryan's procedure, p. 306

Follow through the steps below.

● *Analysis is appropriate when the overall χ_r^2 is significant.*

● *There are a possible $a(a - 1)/2$ pairwise comparisons that may be made.*

A. COMPUTING FORMULA

Critical difference $= \sqrt{[\chi_r^2\alpha]\,[a(a + 1)/6n]}$
where $\chi_r^2\alpha =$ the critical value for χ_r^2 required for significance
            in the overall analysis (Step #7, Friedman test),
    $a =$ the number of levels, and
    $n =$ the number of *subjects*.
Any difference between mean sum of ranks that exceeds the
critical difference indicates a significant difference between groups.

B. DESCRIPTION OF THE STUDY

We shall employ the example of the mock jury that was used to
illustrate the Friedman test. Refer back to that section for the
procedural details. The sums of ranks for the four tests were
$T_1 = 33.5, T_2 = 21.5, T_3 = 31.5$, and $T_4 = 13.5$. There were
ten subjects, and the critical value for χ^2 with $\alpha = .05$ and $a = 4$
was 7.81. The overall χ_r^2 was significant.

C. RECORDING THE DATA

1. The data should be recorded in the same table used to compute
   the overall χ_r^2.

D. COMPUTATIONS

1. Find $\chi_r^2\alpha$.
   Obtain the critical value of χ_r^2 required for significance in the

overall analysis (see Step # 7, Friedman test). This is $\chi_r^2\alpha$.

$$\boxed{\chi_r^2\alpha = 7.81}$$

2. Find $a(a + 1)/6n$.
   Solve algebraically. a is the number of levels in the study, and n is the number of subjects.

$$\boxed{4(4 + 1)/6(10) = 20/60 = .333}$$

3. Find $\sqrt{[\chi_r^2\alpha]\,[a(a + 1)/6n]}$.
   Multiply: # 3 $\times$ # 4. Find the square root of this value.

$$\boxed{\sqrt{(7.81)(.333)} = \sqrt{2.60} = 1.66}$$

4. Compute the *mean* sum of ranks for each group; that is, divide the sum of ranks by n, the number of subjects in each condition.

> Mean sum of ranks:
>
> $$\bar{X}_{T_1} = \frac{33.5}{10} = 3.35,\ \bar{X}_{T_2} = \frac{21.5}{10} = 2.15,$$
>
> $$\bar{X}_{T_3} = \frac{31.5}{10} = 3.15,\ \bar{X}_{T_4} = \frac{13.5}{10} = 1.35$$

5. Order the *mean* sum of ranks from large to small and obtain the difference in mean sum of ranks for all possible pairs.
   a. If you have four or more levels of a, you may find it convenient to summarize the comparisons in a table (see Table 10.6). Prepare the table with the mean sum of ranks arranged in ascending order along the top (column heads) and along the sides (row heads) of the table. In the body of the table (above the diagonal), *list the differences in mean sum of ranks.*
   b. If you have only three levels of a, a less formal procedure for getting difference scores may be appropriate.
6. Any difference between mean sum of ranks that exceeds the critical value, # 3, is a significant difference.

Table 10.6

	$\bar{X}_{T_4}$ 1.35	$\bar{X}_{T_2}$ 2.15	$\bar{X}_{T_3}$ 3.15	$\bar{X}_{T_1}$ 3.35
$\bar{X}_{T_4}$ 1.35	—	.80	1.80*	2.00*
$\bar{X}_{T_2}$ 2.15		—	1.00	1.20
$\bar{X}_{T_3}$ 3.15			—	.20
$\bar{X}_{T_1}$ 3.35				

NOTE: Each value in the body of the table represents the difference between the column and row values.
*$p < .05$.

$$T_1 - T_4 = 3.35 - 1.35 = 2.00 > 1.66, \text{ significant}$$
$$T_3 - T_4 = 3.15 - 1.35 = 1.80 > 1.66, \text{ significant}$$

E. INTERPRETATION

(Remember that low scores and ranks indicated innocence.) The defendant was rated significantly more innocent at the end of the trial (T_4) than at the beginning of the trial (T_1). He was also rated significantly more innocent at the end of the trial than at the end of the third summary (T_3). None of the other differences were significant.

C. SCORE DATA

Because of the popularity and widespread use of the analysis of variance and the consequent interest that statisticians have taken in this test, a large number of specific-comparison tests for use with it have been developed. In addition to procedures that permit the testing of differences between a pair of means or all possible pairs of means, techniques have been worked out that allow you to determine whether all treatment groups differ from a control group, whether treatment means form a straight line (linear trend) or have a significant curvilinear trend, and so on. Only methods for testing the differences between pairs of means or sets of means are covered in this

book. Brief, clear coverage of the other methods can be found in
Edwards (1968), Chapter 8.

For score data, there are a number of different tests that can be
used to determine differences between various pairs of means. Each
of these tests has a number of articulate supporters. We have presented
two tests: Tukey's (a) and Newman-Keuls'. We have also included an
extension of Tukey's (a) test for use with unconfounded means from
interaction tables, which increases the power (the probability of finding a difference when it does exist) of the test.

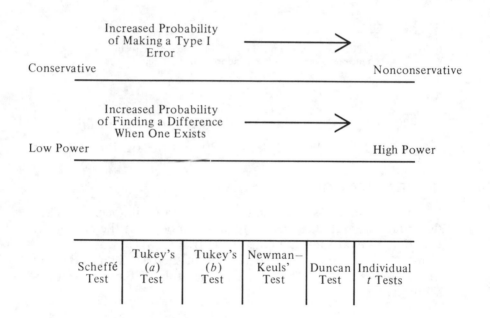

Figure 10.1. *The relationship among conservativeness and
power and several popular multiple-comparison techniques.*

Of these two tests, Tukey's (a) test is the more conservative. It
is also *computationally extremely simple*, since it uses the values
already obtained for the analysis of variance. The complete analysis

should take about five minutes. The Newman-Keuls' test is less conservative (gives less protection against Type I errors), is computationally more difficult, but is more powerful (gives greater protection against Type II errors) than Tukey's (a) test. The relationship between these tests and several other popular specific-comparison techniques is shown in Figure 10.1.

The first line shows how conservative—the second line, how powerful—these tests are relative to each other. The bottom line shows the *order* of the most popular tests on these continua.

Tests of unconfounded means (interaction tables). We also include a test for determining differences between all *unconfounded* means from an interaction table (means that come from the same columns or the same rows in the interaction table). While the usual test of the differences would be based on the total number of comparisons that may be made between all means in an interaction table, this test is made more powerful by adjusting the number of comparisons to reflect only the unconfounded comparisons in the table. The details for this correction are presented in the section for *Tukey's (a) test: Unconfounded Means.* Especially with large interaction tables, the test for unconfounded means is more powerful, on the average, than Newman-Keuls'. It must be emphasized, however, that this test is appropriate only if you wish to make comparisons between unconfounded means. If you wish to make any confounded comparisons, you may *not* use this method. (Such comparisons, when they can be justified, must be made by the method indicated for the regular Tukey's (a) test.)

Must F be significant? A much debated issue is whether the use of specific-comparison tests is legitimate if the overall F ratio is not significant. This question is actually the question about error rates in a somewhat different guise. Some statisticians consider the use of specific-comparison tests after obtaining a nonsignificant F to be illegitimate. You must make your own judgment about your willingness to risk a Type I error. If you decide to run a specific comparison following a nonsignificant F, you may be assured that your decision is supported by at least some statisticians.

Description:

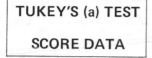

Allows pairwise comparisons of
    means from ANOVA
Use for *differential main effects*
Use Tukey's (a): unconfounded
    means for means from an inter-
    action table
Extremely simple
Conservative test: α controlled
    experimentwise
Relatively low power
Alternative test: Newman-Keuls'
Between-subjects, within-subjects,
    and mixed designs

Follow through the steps below.

● *Have you read the introductory section of this chapter and the
introduction for score data?*

● *Are all relevant assumptions met for your study? If not, do not
continue.*

● *This test is used when you wish to determine whether there are*
differential main effects.
    a. *It may be used to compare all means in a single-factor design.*
    b. *A series of Tukey's tests may be performed, permitting you to
        compare the levels of Factor A, Factor B, and so on, successively.*
    c. *Appropriate for between-subjects, within-subjects, and mixed
        designs, providing the correct MS_{Error} is identified.*
    d. *Use to compare cells of an interaction table only after checking
        Tukey's (a) test: Unconfounded Means, p. 319.*

A. COMPUTATIONAL FORMULA[5]

Critical value (means) $= q_k \sqrt{MS_{Error}/n}$
where q_k is obtained from Appendix 7,
        n = number of observations for each mean compared, and
    MS_{Error} = appropriate error term from ANOVA summary
        table.

---

[5]Should you prefer to perform this analysis using *totals* rather than *means*, you may
simply use totals and the following formula: Critical value (totals $= q_k \sqrt{nMS_{Error}}$. All of
the steps are identical (if total is always substituted for mean), the conclusions are identical,
and, of course, the identical interpretations apply to the totals.

Test $k(k-1)/2$ differences between means against this value.

B. DESCRIPTION OF THE STUDY

Dr. Parker wished to know the amount of social behavior of five species of primates during comparable periods of activity. The amount of touching and contact between individuals of the same species was recorded for 30-minute time blocks over a period of two weeks. An average-social-contact score was obtained for each individual. The mean scores for baboon, chimpanzee, guenon, squirrel monkey, and tree shrew were 22, 20, 11, 7, and 4, respectively. There were seven individuals of each species. A significant F was obtained for the species factor. The summary table for the ANOVA is included for reference (see Table 10.7).

Table 10.7. Summary Table, ANOVA

Source	df	MS	F
Species	4	52.68	4.21
Error	30	12.52	
Total	34		

C. COMPUTATIONS

If you are comparing main-effect means, perform the following analysis for the levels of Factor A before beginning the analysis for the levels of Factor B, C, and so on. If you are comparing means for an interaction table, consider all the means together.
1. From your ANOVA summary table:
   a. Determine the MS_{Error}. In a one-way design with only one error term, there can be no confusion. In two-way or higher-order designs, this value will always be the error term (denominator) in the overall F ratio comparing the means you are testing.[6]

$$MS_{\text{Error}} = 12.52$$

[6] Performing an analysis on the cells of an interaction table in which one factor is within-subjects and the other is between-subjects creates some problems in determining an error term. Any mean, when compared with any other mean, involves a between-subjects comparison or a within-subjects comparison. All between-subjects comparisons must be tested using the between-subjects error term. All within-subjects comparisons must be tested using the appropriate within-subjects error term. Testing both kinds of comparisons with the same critical value always produces an error.

b. Determine n, the number of scores on which each mean is based.

$$\boxed{n = 7}$$

2. Find $\sqrt{MS_{Error}/n}$ by dividing MS_{Error} by n and taking the square root of that value.

$$\boxed{\sqrt{12.52/7} = \sqrt{1.7886} = 1.34}$$

3. From Appendix 7, find q_k
   where df for MS_{Error} is obtained from the ANOVA table,
   $\quad k =$ the number of means that are being compared (this
   $\qquad$ will be a for Factor A, b for Factor B, and so on),
   $\qquad$ and
   $\quad \alpha = .05$

$$\boxed{k = 5, df = 30, \alpha = .05, q_k = 4.10}$$

4. Find your critical value (mean) $= q_k \sqrt{MS_{Error}/n}$ by multiplying # 2 by # 3.

$$\boxed{4.10 \times 1.34 = 5.49}$$

5. You may compare each mean with every other mean. Therefore you have a total of $k(k-1)/2$ comparisons to make.
   a. If you have four or more means, consider preparing a table with the means arranged in ascending order along the top (column heads) and along the sides (row heads) of the table (see Table 10.8). In the body of the table, list the differences between the means. There are $k(k-1)/2$ differences.
   b. If you have only three means, or a clear pattern of differences, a less formal procedure for getting difference scores may be appropriate.
6. Any difference between means (# 5) that exceeds the critical value (# 4) is a significant difference. Mark significant differences with an asterisk in your table.

Table 10.8. Means and Mean Differences

	Baboon (22)	Chimpanzee (20)	Guenon (11)	Squirrel Monkey (7)	Tree Shrew (4)
Baboon (22)	—	2	11*	15*	18*
Chimpanzee (20)		—	9*	13*	16*
Guenon (11)			—	4	7*
Squirrel Monkey (7)				—	3
Tree Shrew (4)					—

*$p < .05$.
NOTE: Each value in the body of the table represents the difference between the column and row values.

D. INTERPRETATION

Both the baboon and the chimpanzee differ significantly from each of the other three primates but do not differ from each other. In addition, the guenon differs from the tree shrew. No other differences were significant. (Compare this pattern of results with that found by the less conservative Newman-Keuls' test, p. 327.)

Description:

Allows pairwise comparisons of all unconfounded means from an interaction table

Relatively powerful

Between-subjects and within-subjects designs

ns for each mean must be equal

TUKEY'S (a) TEST:

UNCONFOUNDED MEANS[7]

---

[7]This method is suggested by Cicchetti (1972).

Follow through the steps below.

● *Have you read the introductory section of this chapter and the introduction for score data, p. 313?*

● *Are all the relevant assumptions met for your study? If not, do not continue.*

● *This test is appropriate when you wish to determine whether there are significant differences among* unconfounded *means from an interaction table. This implies that any single comparison involves only values from a single row or from a single column.[8] If you wish to make comparisons between any* confounded *means in your interaction table, you should use instead the method indicated for Tukey's (a) test, p. 316. (A confounded pair of means comes from neither the same row nor the same column in the interaction table.)*

● *This test may be used to compare unconfounded means in any between-subjects or within-subjects design, providing the correct MS_{Error} or MS_{Errors} is employed. (Tests of means from mixed-design interaction tables provide a more complex case and are not covered in this book.)*

A. COMPUTATIONAL FORMULA (See Footnote 5)

Critical value (means) $= q_{k'} \sqrt{MS_{Error}/n}$
where $q_{k'}$ is obtained from Appendix 7,
  $n =$ the number of observations for each mean compared, and
  $MS_{Error} =$ appropriate error term from the ANOVA summary table.

B. DESCRIPTION OF THE STUDY

We shall employ the example already used for the two-way ANOVA, p. 145. Refer back to that section for experimental details. The Summary Table for this ANOVA is repeated for your convenience (see Table 10.9).

C. INITIAL STEPS

1. *Interaction table.* Compute means for each of the cells of the interaction table, if you have not already done so. See Table 10.10.

[8]These values are unconfounded because systematic differences between them are due to only one factor. For any other comparisons in the interaction table, differences between means may be attributable to differences in two or more variables.

Table 10.9. Summary Table: Two-way ANOVA (2 × 3)

Source	df	SS	MS	F
Test Format (*A*)	2	39.39	19.70	4.15
Gender (*B*)	1	17.36	17.36	3.65
Test Format × Gender (*A* × *B*)	2	48.39	24.20	5.09
Error	30	142.50	4.75	
Total	35	247.64		

Table 10.10

		A_1	A A_2	A_3
			Test Format	
		Control	Consumer	Baseball
	B_1 Boys	3.16	3.50	8.00
Gender: *B*				
	B_2 Girls	3.50	3.67	3.33

2. a. *Determining the number of unconfounded comparisons.* For convenience, call the column variable *A* and the row variable *B*. Any mean may be compared with any other means in its row and in its column in the interaction table. It may *not* be compared with any other means. To determine the *number* of unconfounded comparisons, you may use the following formula:

$[a(a - 1)/2] \times b + [b(b - 1)/2] \times a$ for two-factor designs

where *a* = the number of levels of factor *A*, and
  b = the number of levels of factor *B*.
In other words, compute the number of comparisons that can be made in a single row $[a(a - 1)/2]$ and then multiply this value by the number of rows (*b*). Then compute the number of comparisons that can be made in a single column $[b(b - 1)/2]$; multiply this value by the number of columns (*a*). Sum the value for rows and the value for columns.

For $a = 3, b = 2, [3(2)/2] \times 2 + [2(1)/2] \times 3 = 6 + 3 = 9$

b. Enter Appendix 8 with the number of unconfounded comparisons. Find the number of adjusted k' treatments.

$$k' = 5$$

D. COMPUTATIONS

1. From Appendix 7, find $q_{k'}$.
   where df for MS_{Error} are obtained from the ANOVA Summary Table,
   k' ($k' = k$) is obtained in the preceding section, and
   $\alpha = .05$.

$$q_{k'} = 4.10, \text{ for } df = 30, k' = 5, \alpha = .05$$

2. From your ANOVA Summary Table:
   a. Determine the MS_{Error}. In a simple between-subjects factorial, there should be no confusion. However, this value is the error term on which the interaction (the individual means of which are being tested) is tested.

$$MS_{Error} = 4.75$$

   b. Determine n, the number of scores on which each mean is based.

$$n = 6$$

3. Find $\sqrt{MS_{Error}/n}$ by dividing MS_{Error} by n and taking the square root of that value.

$$\sqrt{4.75/6} = \sqrt{.7916} = .89$$

4. Find your critical value (mean) $= q_{k'} \sqrt{MS_{Error}/n}$ by multiplying #1 by #3.

$$4.10\,(.89) = 3.64$$

5. a. This critical value may be used to test all unconfounded means. Take the first row. Make the $a(a-1)/2$ possible comparisons: $\bar{X}_{A_1 B_1} - \bar{X}_{A_2 B_1}$, $\bar{X}_{A_1 B_1} - \bar{X}_{A_3 B_1}$, $\bar{X}_{A_2 B_1} - \bar{X}_{A_3 B_1}$, and so on. For each pair of means, subtract the smaller mean from the larger. Test each difference against the critical value. Then take the second row; continue the procedure for each of the b rows. Then take the first column. Make the $b(b-1)/2$ possible comparisons. For each pair of means, subtract the smaller mean from the larger. Test each difference against the critical value. Then take the second column; continue the procedure for each of the a columns.
   b. Any difference between means that exceeds the critical value (#4) is a significant difference.

Row 1: $8 - 3.16 = 4.84^*$, $8 - 3.50 = 4.50^*$, $3.50 - 3.16 = .34$

Row 2: $3.67 - 3.50 = .17$, $3.67 - 3.33 = .34$, $3.50 - 3.33 = .17$

Column 1: $3.50 - 3.16 = .34$

Column 2: $3.67 - 3.50 = .17$

Column 3: $8.00 - 3.33 = 4.67^*$

E. INTERPRETATION

Boys answer questions about baseball better than they answer questions on any other subject and better than girls answer questions about baseball.[9] No other differences are significant.

---

[9] Had you wished to ask whether boys answering questions about baseball differed significantly from girls answering questions about consumer problems, it would have been necessary to use the method presented in Tukey's (a) test, p. 316, since this comparison is confounded.

Description:

| NEWMAN-KEULS' TEST |
| SCORE DATA |

Allows pairwise comparisons of
   means from ANOVA
Somewhat complex
Nonconservative test
Relatively powerful
Between-subjects, within-subjects,
   and mixed designs
Alternative test: Tukey's (a) test

Follow through the steps below.

● *Have you read the introductory section of this chapter and the introduction for score data, p. 313?*

● *Are all assumptions met for your study? If not, do not continue.*

● *This test is appropriate when you wish to determine whether there are differences between means.*
   a. It may be used to compare all means in a single-factor design.
   b. A series of Newman-Keuls' tests may be performed to compare the means for Factor A, Factor B, and so on, successively.
   c. Appropriate for testing interactions if all comparisons are to be made. (See Tukey's (a) test: Unconfounded Means *if only unconfounded means are to be tested.)*
   d. Appropriate for between-subject, within-subject, and mixed designs, providing the correct MS_{Error} *is identified.*

A. COMPUTATIONAL FORMULA (See Footnote 5)

Critical value (means) $= q_d \sqrt{MS_{Error}/n}$
where $d =$ the number of ordered means spanned by the comparison,
   q_d is obtained from Appendix 7,
   $n =$ the number of observations for each mean compared, and
   $MS_{Error} =$ the appropriate error term from the ANOVA summary table.
Test ordered mean differences against the appropriate values.

B. DESCRIPTION OF THE STUDY

The study on primates described for the Tukey's (a) test will be used here. Refer back to p. 317 for details.

C. COMPUTATIONS

1. From your ANOVA summary table:
   a. Determine the MS_{Error}. In a one-way design with only one error term, there can be no confusion. In two-way or higher-order designs, this value will always be the error term (denominator) in the overall F ratio comparing the means you are testing.

$$MS_{Error} = 12.52$$

   b. Determine n, the number of scores on which each mean is based.

$$n = 7$$

2. Find $\sqrt{MS_{Error}/n}$ by dividing MS_{Error} by n and taking the square root of that value.

$$\sqrt{MS_{Error}/n} = \sqrt{12.52/7} = \sqrt{1.7886} = 1.34$$

3. From Appendix 7, find your values of q_d
   where df for MS_{Error} is obtained from the ANOVA table,
       $d =$ the number of ordered means spanned by the comparison, and will take on values from 2 to a, the total number of means you are comparing, and
       $\alpha = .05$.

$$df = 30; d = 5, 4, 3, 2; \alpha = .05$$
$$q_5 = 4.10$$
$$q_4 = 3.84$$
$$q_3 = 3.49$$
$$q_2 = 2.89$$

4. Find your critical values (mean) $= q_d \sqrt{MS_{Error}/n}$ by multiplying all the values in #3 by #2. It is probably somewhat more convenient to list these values in descending order.

$$CV_5 = q_5 \sqrt{MS_{\text{Error}}/n} = 4.10 \times 1.34 = 5.49$$
$$CV_4 = q_4 \sqrt{MS_{\text{Error}}/n} = 3.84 \times 1.34 = 5.15$$
$$CV_3 = q_3 \sqrt{MS_{\text{Error}}/n} = 3.49 \times 1.34 = 4.68$$
$$CV_2 = q_2 \sqrt{MS_{\text{Error}}/n} = 2.89 \times 1.34 = 3.87$$

5. Prepare a table of differences between means by arranging the means in descending order along the top (column heads) and along the sides (row heads) of the table. Table 10.11 shows this procedure. In the body of the table, list the differences between the means. There are $a(a - 1)/2$ differences.

Table 10.11. Means and Mean Differences

	Baboon (22)	Chimpanzee (20)	Guenon (11)	Squirrel Monkey (7)	Tree Shrew (4)	
Baboon (22)	—	2	11*	15*	18*	
						$CV_5 = 5.49$
Chimpanzee (20)		—	9*	13*	16*	
						$CV_4 = 5.15$
Guenon (11)			—	4*	7*	
						$CV_3 = 4.68$
Squirrel Monkey (7)				—	3	
						$CV_2 = 3.87$
Tree Shrew (4)					—	

*$p < .05$.

6. Each mean must be tested against the appropriate critical value; however, carrying out the tests in a prescribed order prevents inconsistencies in interpretation.
   a. Every difference between means is tested against the critical value for the number of steps between the means. Adjacent means have a d of 2, means with a single intervening mean have a d of 3, and so on. For the single pair of means with the

greatest range, $d = a$. The means that must be tested with each critical value are indicated in Table 10.11.

b. Tests must be made beginning at the right side of the top row. If the first mean difference is significant, then the second value in the row is tested (with the appropriate critical value), and so on. When a mean difference is found not to be significant, no further tests are made on that row. Tests are then made beginning at the right side of the second row. Tests are made of the mean differences, using the appropriate critical value, until one is found that is not significant or until you come to the column in which there was not a significant difference in the first row.

NOTE: The reason for this restriction can be seen readily by taking a specific example. Suppose that in Table 10.11 you found that the difference between squirrel monkeys and baboons (in the first row) was *not* significant. To find a value in that column to be significant would mean that, although squirrel monkeys and baboons were not different, squirrel monkeys and chimpanzees or squirrel monkeys and guenons, both of which are more similar than squirrel monkeys and baboons, are different. This inconsistency is prevented by the imposition of the general rule that no significant difference may be contained within the range of a nonsignificant difference. If the first value at the right of any row is not significant, the testing ends, for there can be no further significant differences in either rows or columns in the table.

c. Any difference between means that exceeds the appropriate critical values is a significant difference.

D. INTERPRETATION

The baboon and the chimpanzee engaged in significantly more social contact than the guenon, the squirrel monkey, and the tree shrew; but the baboon and the chimpanzee did not differ from each other. The guenon had significantly more social contact than did the squirrel monkey or the tree shrew. The baboon and the chimpanzee did not differ from each other; neither did the squirrel monkey and the tree shrew. (Compare this pattern of results with that found by the more conservative Tukey's (a) test, p. 319.)

CHAPTER ELEVEN

STRENGTH-OF-ASSOCIATION MEASURES

The delightful glow that researchers feel when their data turn out to be "significant" often seems to impair their judgment on the meaning of their findings. When a statistical test indicates that the relationship between two variables is significant, it tells us only that, at a specified probability level, the relationship exists to *some* extent in the population from which the subjects have been randomly drawn and that it is not due to the operation of chance sampling factors. It cannot be stressed too strongly that a statistical test tells us no more. A relationship may be very weak and still be quite real. We'll try to illustrate this fact with a somewhat facetious example.

Somewhere on Madison Avenue there lurks an account executive who is interested in whether blondes really do have more fun than brunettes. He has available a good random sample of 1002 young singles from a local directory. To make the example work, let's assume that he has developed an objective score-data measure of hair color and a test that not only measures how much fun an individual has but also yields score data. He makes both measurements on his sample and correlates the two sets of numbers using a Pearson r. His test statistic yields a value of r equal to $+ .063$ with 1000 df. He opens a statistics book, with trembling fingers, finds the appropriate table, and discovers to his joy that an r of $+ .063$ with 1000 df is indeed significant ($p < .05$).

The research hypothesis that blondes have more fun than brunettes has been supported. There is less than a 5% chance that the actual population correlation is zero. However, a related statistic, the *coefficient of determination*, tells us the proportion of variance in one set of scores that can be accounted for from the other set of scores. We may obtain the coefficient of determination by squaring r. If we square the $r = +.063$ for our study, we obtain a coefficient of determination of .00397. Worded in terms of the study, a little less than 4/1000 of the variance in the fun scores can be attributed to hair color. Expressed in a different way, 99.6% of the variance in the fun scores is attributable to other unspecified factors. Despite the fact that the relationship between hair color and having fun is significant, it is so small as to have no practical significance whatsoever. Although this is an extreme case, the general principle is well worth pondering. Particularly with large samples, very weak relationships may well be significant.

Many people who engage in research are not aware that techniques are available to assess the strength of relationships and experimental effects. Researchers too often assume that very high levels of significance (.01, .001, and so on) indicate a very strong relationship between variables. This assumption is not only philosophically incorrect but pragmatically wrong as well. With very large samples, very small effects can reach extreme levels of statistical significance. It seems clear, however, that intelligent conclusions cannot be drawn from data unless you have some idea of the strength of the relationship you have discovered. We believe that for much research the correct strategy is first to ask, "Is the relationship real?" If the answer is affirmative, then ask, "How strong is the relationship?"

It is often very difficult to make reasonable inferences about the meaning of research results if a strength-of-association measure is not computed. As part of our preparation for this book, we had an assistant calculate strength-of-association measures on most of the published studies in American Psychological Association journals for the year 1964. To our amazement, over 50% of the published studies reported experimental effects that, even in highly controlled experimental situations, accounted for less than 5% of the variance in the dependent variable studied. In every case, the results had reached an acceptable level of statistical significance. In most cases, authors discussed their findings as though they had discovered powerful determiners of human behavior, whereas in fact the strength-of-association measures indicated weak relationships between the experimental treatments and the ensuing behavior. In comparing studies that revealed very strong relationships, as opposed to those with weak relationships, we found no difference in the authors' interpretations of their data.

The moral of this story is that what you say about your data should be based on the strength, as well as on the reality, of the relationship you find. If you can account for less than 5% of the variance in the dependent variable from the independent variable, you should probably point out that, although the relationship may be real, it is quite weak. On the other hand, if you find that you can account for 75% of the variance in your dependent variable on the basis of your independent variable, you are justified in making some strong inferences about the results and in talking as though you have a powerful effect.

One limitation on most strength-of-association measures, however, is that the strength-of-association measure you obtain holds only for the particular situation in which it was discovered. If you determine in your experiment that 50% of the variance in maze-learning scores for a group of rats can be attributed to the number of hours that the rats were food-deprived, you are not justified in concluding that, in general, 50% of the variance in maze-learning scores can be attributed to amount of deprivation. That finding is true only for the particular maze used and for the conditions that prevailed at the time you ran your experiment.[1]

Unfortunately, as yet there are no hard and fast rules to tell you how strong a relationship you need before you can begin to feel happy about your results. A good dose of common sense is probably the best guideline. Judging from the present state of the art in the behavioral sciences, any time you can account for more than 10% of the variance, you are doing better than the vast majority of studies.

One of the most useful features of strength-of-association measures is that they can serve as a guideline to indicate to you how well you understand the phenomenon you are studying. All too often, researchers perform a study, get a significant result, and believe that they have answered the experimental question. A strength-of-association measure indicating that only a weak relationship has been found should serve as an impetus to sharpen your conceptualizations, to refine techniques and methodology, and to run another study, this time exerting more control over extraneous variables. If a second study shows a stronger relationship, you know that you are on the right track, and you can hope that a third study will yield a still stronger relationship. The widespread acceptance of strength-of-association measures should help to encourage more programmatic research. Similarly, if you have refined a particular variable to the point that no further variance may be accounted for, you may add a second factorial variable to see if the percent of variance accounted for may be increased still further by the addition of the new main effect and the interaction. Again, the use of strength-of-association measures should encourage an approach to research that moves from the simple to the more complex and provides clear feedback on the

---

[1] Of course, the amount of variability that is *not controlled* in the study (by the rigor of the experimental controls) strongly affects the amount of variance accounted for by the independent variable. For exactly the same independent variable, the strength of association will be larger in a study in which extreme care is exercised in controlling all extraneous variables than in a study in which a number of extraneous variables are permitted to vary randomly.

success of each step. Both of these trends would be welcome additions to research strategies in the behavioral sciences.

COMPUTATIONS

In general, an estimate of the strength of association in a given research design may be made through the use of the formula

$$\hat{r}_m^2 = \frac{Q^2}{Q^2 + N}$$

where $\hat{r}_m^2$ = estimate of the strength of association,
Q = the numerical value of the inferential statistic, and
N = the sample size.
This formula gives only a rough estimate of the strength of association in the study sample, but for most purposes a rough estimate is better than none. Specific formulas are presented in the following pages. As more statisticians become interested in this problem, refinements in both conceptualization and computation can be expected. The rest of this chapter is devoted to specific applications that have been worked out.

A. FREQUENCY DATA

The appropriate strength-of-association measure for the χ^2 test is the square of the contingency coefficient, C^2. The computational formula for C^2 is

$$C^2 = \frac{\chi^2}{\chi^2 + N} \cdot$$

Unfortunately, C^2 is not an extremely useful strength-of-association measure. Although the theoretical limits of C^2 are from 0 to 1, the practical limit to C^2 is much less than 1. C^2 approaches its upper limit as the number of subjects and the number of levels of the independent variables increase. Thus the strength of association in χ^2s of different size cannot be directly compared. Furthermore, the intuitive meaning of C^2 is not clear. It is not a variance-reduction measure, although it is often interpreted in this way. Nevertheless, it still has limited usefulness for comparing χ^2s of the same size (for example, a number of 2 × 2s) in which the data were gathered under comparable conditions.

Example and interpretation. As an example, let's assume that a researcher gathered data on the number of freshmen and seniors who owned or did not own cars at two local high schools, one in a wealthy area and one in a poor area. In both high schools, he found a relationship between car ownership and class standing, with seniors being more likely to own cars than freshmen. In the poor area, the χ^2 was 4.0, with a sample of 40 students. In the rich area, the χ^2 was 28.3, also with a sample of 40 students. C^2 for the poor school would be $4/(4 + 40) = 4/44 = .091$. C^2 for the wealthy school would be $28.3/(28.3 + 40) = 28.3/68.3 = .414$. Although the relationship between car ownership and class standing is significant at both schools, it is apparent that there is a much stronger relationship at the wealthy school than at the poor one.

B. ORDERED DATA

Little attention has been paid to developing strength-of-association measures for ordered data. For tests such as the sign test, which is evaluated directly from a table and does not yield an inferential statistic, strength of association cannot be estimated at present. For tests that yield an inferential statistic, the general formula on p. 332 must be modified. The appropriate formulas are:

Kruskal-Wallis test: $\qquad \eta^2 = \dfrac{H}{N - 1}$ $\qquad$ (H is not squared because it is already a squared statistic.)

Friedman test: $\qquad \eta^2 = \dfrac{X_r^2}{N - 1}$ $\qquad$ (N = the number of subjects times the number of observations per subject.)

Rank-sums test: $\qquad \eta^2 = \dfrac{Z^2}{N - 1}$ $\qquad$ (This formula can also be used with other tests for ordered data that yield Z scores when the sample size becomes large.)

Example and interpretation. Let's assume that a researcher has tested 18 blind children and 12 sighted children on their ability to make fine auditory discriminations. Each child is ranked and a rank-sums test is performed that yields a Z of $+ 3.8$, indicating that the blind children do significantly better than the sighted ones. η^2 equals $3.8^2/$ $(30 - 1) = 14.44/29 = .498$. Thus 49.8% of the variance in the ranks of auditory ability may be attributed to the presence or absence of sight in these children.

C. SCORE DATA

t test. The appropriate strength-of-association measure for the *t* test is eta squared (η^2). The computational formula is

$$\eta^2 = \frac{t^2}{t^2 + df} .$$

Example and interpretation. A researcher wished to determine whether stimulation of rats during their infancy affected the speed with which they learned a discrimination as adults. One group of ten rats was handled daily from the time they were weaned until they were adults. The second group of ten rats was subjected only to the usual taming procedures preceding the discrimination trials. The rats handled as infants learned the discrimination with a mean of 9.5 trials. The un-handled rats required 15.4 trials to learn the discrimination, $t = 4.65$ with $df = 18$. η^2 for these values equals $(4.65)^2/(4.65^2 + 18)$ $= 21.63/39.63 = .55$. We have a very strong experimental effect. In this sample, 55% of the variance is accounted for by factors associated with the handling of these subjects.

ANOVA. Since analysis of variance is the most widely used statistical test, it is not surprising that most of the work on strength-of-association measures relates to this test. And, as is often the case, the more an area is explored, the more complex it turns out to be. We focus our attention on two of the simpler strength-of-association measures.

    A rough estimate of the strength of association in a fixed-effects analysis of variance can be obtained through the use of the correlation ratio, η^2. The computational formula for η^2 is

$$\eta^2 = \frac{SS_x}{SS_{Total}}$$

where SS_x = the sum of squares of any treatment effect or the interaction of any treatment effects.

    This is computationally a very simple formula, since all values may be obtained directly from the analysis-of-variance source table. η^2 gives a reasonably accurate estimate of the strength of association for between-subjects, within-subjects, and mixed designs, and the interpretation is quite straightforward. η^2 represents the proportion of variance in the dependent variable that may be accounted for by the independent variable(s). However, the major disadvantage to the use of η^2 is that it gives an estimate of the strength of association in your *sample*, when you may want an estimate of the strength of association in the *population*. Depending on the design, η^2 may either overestimate or underestimate the strength of association in the population.

A more precise estimate of the strength of association in the population may be obtained by the use of omega squared (ω^2). Mathematically, ω^2 is a ratio of the expected mean square (*EMS*) for the treatment effect under consideration to the sum of the expected mean squares for all effects. (Don't confuse expected mean squares with the actual mean squares that are obtained from an analysis of variance. They are not the same. Expected mean squares are equations that express a particular variance component as a function of population-variance components.) For an analysis of variance having two independent variables A and B, ω^2 for the strength of association between independent variable A and the dependent variable can be estimated by the formula

$$\omega_A^2 = \frac{SS_A - (df_A)(MS_{Error})}{MS_{Error} + SS_{Total}}.$$

This formula is appropriate only when all independent variables (if there are more than one) are factorially combined, there are equal numbers of subjects in each cell, all groups of subjects are independent, and the fixed-effects model is assumed. Similarly, ω^2 for B and $A \times B$ can be estimated by substituting the appropriate SSs and dfs in the numerator of the formula.

Example and interpretation. To illustrate the use of η^2 and ω^2, we have presented a hypothetical source table for a 2 $\times$ 3 analysis of variance (see Table 11.1).

Table 11.1. Source Table: 2 $\times$ 3 Analysis of Variance

Source	df	SS	MS	F	p
A	1	6.14	6.14	1.60	n.s.
B	2	40.30	20.15	5.25*	< .05
A × B	2	30.60	15.30	3.98*	< .05
Error	54	207.36	3.84		
Total	59	284.40			

*$p < .05$.

Since the main effect of A is not significant, we shall perform ω^2 and η^2 only on the main effect of B and on the interaction of A and B.

$$\omega_B^2 = \frac{40.30 - 2(3.84)}{3.84 + 284.40} = \frac{32.62}{288.24} = .113$$

$$\omega_{A\times B}^2 = \frac{30.60 - 2(3.84)}{3.84 + 284.40} = \frac{22.92}{288.24} = .080$$

$$\eta_B^2 = \frac{40.30}{284.40} = .142$$

$$\eta_{A\times B}^2 = \frac{30.60}{284.40} = .108$$

Two things are apparent from our example. First, although the effects of both B and $A \times B$ are significant, neither accounts for a great deal of variance in the dependent variable. In both cases, we have a rather weak experimental effect. Second, η^2 gives us a slightly higher estimate than ω^2. η^2 (which indicates the variance accounted for in the sample) always yields a slightly higher estimate than ω^2 (which estimates the population variance accounted for), since the population variance accounted for is assumed to be always slightly smaller.

Whenever the four assumptions above are not met—that is, you have a random-effects model, repeated measures, or mixed designs—estimations of population variance accounted for become quite complex. Nevertheless, the strength of association between independent and dependent variables may still be estimated by using the ratio of the expected mean square for the particular treatment effect to the sum of the expected mean squares for all of the effects.[2]

We strongly urge the computation of ω^2 whenever analysis of variance is the statistical test employed. For the majority of research involving the fixed-effect model and simple factorial designs on independent groups, the simplified computing formula we have given is appropriate. If your design does not fit this framework, go to the sources that we have recommended and perform the appropriate computations to obtain the expected mean squares. If you find such computations too difficult or tedious for your design, then η^2 should be computed to give at least an estimate of the strength of association involved in your study. The interpretations of η^2 and ω^2 are identical. with the exception that η^2 gives an estimate of the amount of variance in the dependent variable that can be accounted for by the independent

---

[2] See Winer (1962), p. 151, for principles involved in obtaining expected mean squares; see Vaughn and Corballis (1969) for worked examples of many complex designs and principles for estimating strength of experimental effects.

variable in the *samples* that have been studied, whereas ω^2 gives an estimate of the amount of variance that can be accounted for in the *population.*

It should be pointed out that even in the case of ω^2, information is lacking. It is not known whether ω^2 is a biased estimate of population strength of association or, if so, in what direction it is biased. However, this type of information will eventually become available; in the meantime, even without this information, strength-of-association measures are extremely useful tools in our statistical workshop.

Kind of Data

Score	Ordered	Frequency	Correlation Coefficient
2			Pearson $r_{x,y}$, p. 347
	2		Spearman ρ (rho), p. 352
		2	C, p. 356
1	1		Spearman ρ (rho), p. 352
1		1	η (eta), p. 358
	1	1	η (eta), p. 361

CORRELATION COEFFICIENTS AND SIGNIFICANCE TESTS FOR CORRELATION COEFFICIENTS

LIMITATIONS AND EXCEPTIONS

1. Correlational techniques presented in this chapter are applicable only for relationships between two variables. Multiple correlations, partial correlations, and other more complex correlational techniques are not presented.
2. You must have pairs of measurements for each subject, or paired measurements for subjects that form logical pairs.
3. No test is provided for score/order data. We recommend that you convert the score data to ordered data and perform the appropriate test for pairs of ordered measurements.

AN OVERVIEW

A correlation is an interrelationship between two (or more) variables. A researcher generally chooses a correlation coefficient in order to summarize the strength of a relationship between variables with a single number.

The correlation techniques presented in this chapter are applicable only when the relationship of interest involves *two* variables.

There are, however, a variety of sophisticated techniques that permit you to evaluate the effects of more than two variables.[1] Our decision to omit these more advanced topics was purely pragmatic. Their use is relatively infrequent among beginning researchers, and they are both computationally difficult and highly specialized.

If you compare the branching programs for tests of significance with the branching program for correlation coefficients, you will see that fewer factors must be considered in selecting a correlation statistic. There are some similarities between the perspectives required for considering correlation coefficients and those required for considering significance tests for frequency data (see Chapter 5). In both cases, we find that the distinction between independent variables and dependent variables loses its usefulness. In the case of correlations, that otherwise important distinction becomes blurred because correlations can express the strength of a relationship between two dependent variables (a conclusion never found in significance testing) or between an independent variable and a dependent variable. We always have two measurements for each subject, but we need not ask whether the measures have the status of dependent or independent variables.

One consequence of this blurred distinction between dependent and independent variables is that our admonition from the chapters on significance testing—that two different dependent variables never appear in a single analysis—does not hold for correlations. Earlier, we used the example of a study of a group of rats for which we had obtained both weights and number of trials to learn a maze. We cannot ask the test of significance question: does the weight of the animals differ from the number of trials they required to learn a maze? That is, we cannot perform a test of significance that asks whether these two dependent variables are different. The question simply doesn't make sense. The numbers are not equivalent, and, of course, we already know that weight and time to learn are inherently different. Therefore these two dependent variables would never appear in a single test of significance.

We might expect that light animals would run more quickly than heavier animals and thus learn the maze more rapidly. To ask this question is quite legitimate. The question may be rephrased: are weights and trials to learn the maze related? The lack of equivalence of the measurements that disturbed us in significance testing is no

---

[1] If there are more than two variables involved in the relationship, a more complex correlational procedure, such as multiple regression or partial correlation, is required. See Harshbarger (1971), Chapters 15, 16, 17, and 18, for an introduction to these topics.

longer important; neither does it matter that weight and time to learn are inherently different.

Although we need not be concerned with the status of the variables (dependent or independent variables), we must be concerned with the kind of data we have obtained. Correlations may be performed on score data, ordered data, frequency data, or any combination of these data types. The appropriate analysis for any set of data depends on the kind of data we have.

All correlation coefficients require paired measurements. Each subject[2] must have a score, rank, or category (in the case of frequency data) for each variable. More specifically, the correlation coefficients may be used when (1) there are two sets of scores (each subject has a score for each variable), (2) there are two sets of ranks (each subject has a rank for each variable), (3) there are two sets of categories (each subject is classified in a category for each variable), (4) there are a set of scores and a set of ranks (each subject has a score on one variable and is ranked on the other variable), (5) there are a set of scores and a set of categories (each subject has a score on one variable and is categorized on the other variable), or (6) there are a set of ranks and a set of categories (each subject is ranked on one variable and categorized on the other variable).

Although a variety of correlations may be used, we have presented the simplest, most straightforward procedures for answering these correlation questions. For two sets of score data, we have presented the Pearson r, which is by far the most common correlational technique. For two sets of ranked data, we have presented the Spearman rho (ρ). Many statisticians[3] believe that Kendall's tau (τ) or Goodman and Kruskal's gamma (γ) are conceptually superior to ρ, but τ and γ are considerably more complex computationally, and correction procedures must be applied when ranks are tied. For two sets of categories, the contingency coefficient, C, is presented.

When score data are to be related to ordered data (ranks), the multiserial correlation coefficient, r_{ms}, is often used.[4] We do not present this statistic, however, because of the complex computations involved in its use. We recommend instead that the beginning researcher convert the score data to ordered data and compute a

---

[2] Sometimes measurements on other logical pairs, such as twins or husband-wife combinations, are correlated. In such cases, this sentence should read: "Each subject in each pair must have a score. . ."

[3] See Hays (1963), pp. 647–656, for a discussion and comparison of tau (τ) and gamma (γ).

[4] If you prefer to use the multiserial correlation coefficient, see Harshbarger (1971), p. 391.

Spearman ρ. For a set of scores and a set of categories, or for a set of ranks and a set of categories, we present the correlation ratio, eta (η).

TESTING FOR SIGNIFICANCE OF CORRELATIONS

A correlation coefficient indicates the strength of a relationship. It does not, however, indicate whether the relationship obtained differs significantly from zero. For that reason, each correlation must be followed by a test of significance. In each case, the simplest null hypothesis for the significance test is that the population correlation equals zero. (As usual, a variety of other null hypotheses may also be tested.) If the null hypothesis is rejected, the researcher may conclude that the relationship in his sample is representative of a real relationship in the population from which the sample was drawn. When a correlation coefficient is reported, the results of the test of significance should also be reported.

The significance test, happily, requires little or no additional work. To determine the significance of r and rho (ρ), you enter a table that requires only the value of the statistic and the sample size. Since the value of those coefficients is derived computationally from the results of the associated significance tests, the significance of C and eta (η) is known before the coefficients are computed.

INTERPRETING CORRELATIONS

In interpreting correlations, one caution is imperative. When you find evidence for the existence of a relationship, you have not found evidence that one factor has "caused" the other. In our earlier example, a significant correlation between our rats' weight and their running speed does not mean that rats' running faster is caused by their being lighter; neither does it mean that the greater speed caused the lesser weight. In many cases, both factors are themselves caused by yet a third variable. In the present example, weight and speed might be intrinsically related to degree of hunger or to the age of the animals. In general, causative statements are inappropriate in the interpretation of correlations and should be avoided.

If the researcher's main aim is to establish whether or not a relationship exists in the population, the significance test for the correlation coefficient answers that question. The researcher, however, usually wants to know how strong the relationship is. Two coefficients, the Pearson r and the Spearman ρ, have limits of $+1$ to -1. For a value of $+1$, the relationship is both perfect and positive (high scores or ranks for one variable are related to high scores or ranks for the other).

For a value of −1, the relationship is perfect but negative (high scores or ranks for one variable are associated with low scores or ranks for the other). The sign always indicates the direction of the association. The numerical value indicates the strength. The closer the value is to 1, the stronger the relationship; the closer it is to 0, the weaker the relationship—regardless of sign. A correlation of 0 indicates that there is no relationship at all.

In contrast, the other two coefficients, η and C, have no sign and have limits of 0 and 1. Why should these coefficients have no sign? You will remember that both of these coefficients involve frequency data. Since categories ordinarily have no mathematical properties,[5] it makes no sense to talk about positive or negative relationships. That is, since categories such as gender, political party, or living accommodations do not ordinarily have clear-cut high and low values, to describe an association as being between high values of one variable and low values of the other is meaningless. η has a fairly broad range and can assume the value of 1 when there is a perfect relationship between variables. For C, on the other hand, the theoretical limit of 1 is virtually impossible to reach even when the relationship is perfect.

Although it is relatively easy to describe verbally a perfect relationship or no relationship at all, it is quite difficult to describe the meaning of a Pearson r of .36 or an η of .54. In general, there are two things you can do to help yourself interpret a correlation coefficient. The first is to plot the two sets of data and visually inspect the relationship between them. As you compare different sets of data having relationships of different strength, you gradually sense the meaning of the correlation coefficients. The second useful aid is to square the coefficients to obtain a strength-of-association measure (this usage is discussed in detail in the sections below). It is often easier to understand a Pearson r of +.71, for example, by realizing that about half of the variance ($r^2 = .49$) in one set of scores can be accounted for by another set of scores than it is to understand intuitively the degree of relationship expressed by the value +.71.

---

[5] Some categories do have mathematical properties. For example, an experiment can test the effects of 24, 48, and 72 hours of sleep loss on motor performance by assigning ten subjects to each condition. Or subjects may be classified as above or below the median on one test and above or below the median on a second test. An η obtained in the first case or a C obtained in the second case would be positive, even though the relationship might be negative (greater sleep loss was associated with lower performance scores; subjects scoring above the median on the first test tended to score below it on the second). In such cases, the researcher must determine the direction of the relationship by inspection, since the statistic does not tell him.

CORRELATION AND PREDICTION

There is a close relationship between correlation and prediction. The stronger the relationship between two variables, the more accurately one can predict one variable from knowledge of the other. Unfortunately, knowledge of the value of a correlation coefficient does not always give precise information about the predictability of one variable from the other. For some correlation coefficients, such as the Pearson r, there is a direct relationship between the value of the coefficient and the degree of error reduction in the prediction. For others, such as η and C, the relationship is less clear. Several simple examples illustrate these points.

Coefficient of determination. Let's assume that a Pearson r performed on two sets of test scores has yielded a correlation coefficient of +.90. The Pearson r may be squared to yield the *coefficient of determination*. In this case, the coefficient of determination is .81. Figure 12.1 shows a scattergram representing such a relationship with a mathematically determined line of best fit for predicting X from Y. To predict a score on test X from a score on test Y, run a line parallel to the baseline, from the predictor score to the line of best fit, and then drop a perpendicular line to the baseline. In the example, if a score of 40 was obtained on test Y, you would predict a score of 19.5 on test X. Without information about the relationship of Y and X, the best prediction of any S's score would be the mean of the X distribution. The coefficient of determination indicates that knowledge of the Y score and of the relationship of Y to X allows us to predict Ss' scores in such a way that we can account for 81% of the total variance in the Ss' scores, thus greatly reducing our error of prediction. Furthermore, the relationship is symmetric. Using the X scores to predict the Y scores also reduces error of prediction by the same amount. In this case, the line of best fit for predicting Y from X would be used.

In general, predicting from ρ is very similar to predicting from r. In fact, except when there are ties, ρ (for data originally measured in ranks) has the same value that r would have.

In the case of η (categories and scores), the relationship is asymmetric. Using a case that unambiguously illustrates this assymmetry, let's assume that we wished to determine the relationship between political party affiliation (Democrat or Republican) and political liberalism (as measured by a test that yields score data). Five Democrats and five Republicans were given the test. The five Democrats scored 22, 20, 18, 16, and 12. The five Republicans scored 10,

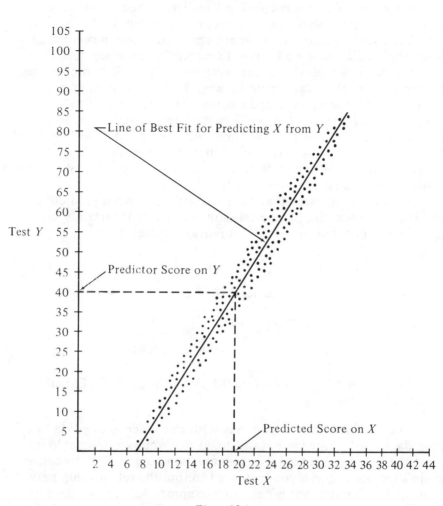

Figure 12.1

8, 6, 4, and 3. If we were asked to predict the score of a subject taking the test without knowing the political party affiliation of any of the subjects, we would predict the mean liberalism score for the ten subjects. However, if we knew which political group the subject came from, we would predict the mean political liberalism of that group rather than the overall mean. Computing η^2 indicates that knowledge of political party affiliation accounts for 71.4% of the variance in the liberalism scores. However, in predicting from liberalism scores to political party, we would have no error of prediction at all in this case, since we would always predict Democrat when the score was above 11 and Republican when the score was below 11. Thus η^2 does not yield accurate information about error reduction in prediction when the prediction is made from the scores to the categories.

Our final case illustrates the assymmetry of C. Suppose we are concerned with the relationship between being frustrated or not frustrated and passing or failing a particular test. C computed on our data equals .69, and C^2, the coefficient of determination, is .47. It would seem reasonable that knowing whether a person was frustrated or not would reduce our error of prediction somewhat. Yet in this case our predictions would be perfect, since all people who were frustrated failed and all who were not frustrated passed.

Thus it may be seen that the relationship between prediction and the numerical size of a correlation coefficient is certainly high and positive but that for η and C it is also complex.[6]

	Pass	Fail	
Frustrated	0	20	20
Not Frustrated	20	0	20
	20	20	40

NOTE ON ASSIGNING RANKS AND THE PROBLEM OF TIES

In a number of cases, you may wish to convert score data to rank data. First, you may be concerned that your score data do not meet the assumptions for interval data, and you may wish to convert them to ranks. Second, you may wish to find the relationship between score and rank data; in this case, we recommend that score data be transformed to rank data so that the simpler Spearman ρ may be performed. In either case, see the detailed examples provided on this topic in Chapter 6 (p. 100). Assignment of ranks in case of ties is discussed in the same section.

NOTE ON THE ANALYSES AND WORKED EXAMPLES

1. Each analysis is presented with step-by-step, spelled-out procedures. There is a worked example for each analysis. The example (1)

[6] Chapters 15, 16, 17, and 18 in Harshbarger (1971) provide a good introduction to the problems of prediction from correlation coefficients.

begins with a description of the study, (2) presents an appropriate null hypothesis, (3) indicates the steps through the branching program, (4) follows step by step through the computations, and (5) presents the statistical and research outcomes for the study.

2. Since later computations refer back to steps by number ("#4 less #5"), you will save time if you list the products from each step by number as you proceed. In addition, you can locate and correct errors in your computations more easily if you have recorded your intermediate steps systematically.

3. Accompanying each analysis is a brief *Description* that lists in a summary form when the analysis may be used, special requirements, and so on.

4. Each analysis is followed by a significance test for the statistic and interpretation of the correlation.

Description:

PEARSON *r*

Two variables
Both score data
One score on each variable for each subject

Follow through the steps below.

- *Verbal equivalents for all notation are presented in the* Computations *section below.*

- *If you are experienced with computations for the Pearson* r, *go directly to the computing formula below. If you require step-by-step guidance, begin with Section B,* The Example.

A. COMPUTING FORMULA

$$r_{x,y} = \frac{N\Sigma XY - (\Sigma X)(\Sigma Y)}{\sqrt{[N\Sigma X^2 - (\Sigma X)^2][N\Sigma Y^2 - (\Sigma Y)^2]}}$$

where N = the number of paired scores,
$\quad X$ = each raw score on the first variable, and
$\quad Y$ = each raw score on the second variable.

B. THE EXAMPLE

1. After extensive navigational training, each Air Force trainee was given the Navigational Aptitude Test (NAT) to determine his navigational skills. The trainers wished to determine how closely the trainees' scores on the NAT were related to a measure of the trainees' spatial-visualization abilities. For the latest group of ten recruits, the Spatial Visualization Test (SVT) was administered just prior to the beginning of training.
2. *Null Hypothesis*
   a. *General.* The observations are drawn from a population in which the correlation between the variables is zero.
   b. *Specific.* In the population being sampled, the correlation between scores on the Navigational Aptitude Test and scores on the Spatial Visualization Test is zero.
3. *Steps through the Branching Program for This Example*
   a. *Statistical technique needed:* **correlation coefficient**
   b. *Kind of data:* **variable 1, score data:** each subject has a score on the Navigational Aptitude Test. **Variable 2, score data:** each subject has a score on the Spatial Visualization Test.

C. RECORDING THE DATA

Each subject has two scores, one for each variable. Record the data in a table (see Table 12.1) having the following three headings:
1. *Subjects.* List each subject by number or by some other identifying rubric.
2. *X scores.* List the score for each subject on the first variable.
3. *Y scores.* List the corresponding score for each subject on the second variable.

NOTE: If calculations are to be carried out by hand rather than on a calculator, add three more headings to the table, labeled X^2, Y^2, and XY.

Table 12.1

Recruit	SVT Score X	NAT Score Y	X^2	Y^2	XY
B.R.	7	9	49	81	63
R.S.	6	8	36	64	48
A.M.	9	11	81	121	99
C.B.	3	5	9	25	15
P.O.	6	5	36	25	30
M.T.	10	9	100	81	90
L.T.	4	4	16	16	16
R.L.	2	1	4	1	2
F.K.	7	8	49	64	56
G.J.	5	5	25	25	25
Totals	59	65	405	503	444

D. COMPUTATIONS

1. Obtain N.
   Count the number of paired scores.

$$\boxed{N = 10}$$

2. Obtain ΣX.
   Sum all of the X scores.

$$\boxed{7 + 6 + \cdots + 5 = 59}$$

3. Obtain ΣY.
   Sum all of the Y scores.

$$\boxed{9 + 8 + \cdots + 5 = 65}$$

4. Obtain ΣX^2.
   Square each X score, then sum.

$$\boxed{49 + 36 + \cdots + 25 = 405}$$

5. Obtain ΣY^2.
   Square each Y score, then sum.

$$81 + 64 + \cdots + 25 = 503$$

6. Obtain ΣXY.
   Multiply each X score by its corresponding Y score, then sum.

$$(7 \times 9) + (6 \times 8) + \cdots + (5 \times 5)$$
$$= 63 + 48 + \cdots + 25 = 444$$

7. Obtain $(\Sigma X)^2$.
   Square the result of #2.

$$59^2 = 3481$$

8. Obtain $(\Sigma Y)^2$.
   Square the result of #3.

$$65^2 = 4225$$

9. Obtain $N\Sigma XY$.
   Multiply: #1 $\times$ #6.

$$(10)(444) = 4440$$

10. Obtain $(\Sigma X)(\Sigma Y)$.
    Multiply: #2 $\times$ #3.

$$(59)(65) = 3835$$

11. Obtain $N\Sigma XY - (\Sigma X)(\Sigma Y)$.
    Subtract: #9 $-$ #10.

$$4440 - 3835 = 605$$

12. Obtain $N\Sigma X^2$.
    Multiply: #1 $\times$ #4.

$$(10)(405) = 4050$$

13. Obtain $N\Sigma X^2 - (\Sigma X)^2$.
    Subtract: #12 − #7.

$$4050 - 3481 = 569$$

14. Obtain $N\Sigma Y^2$.
    Multiply: #1 × #5.

$$(10)(503) = 5030$$

15. Obtain $N\Sigma Y^2 - (\Sigma Y)^2$.
    Subtract: #14 − #8.

$$5030 - 4225 = 805$$

16. Obtain: $\sqrt{[N\Sigma X^2 - (\Sigma X)^2][N\Sigma Y^2 - (\Sigma Y)^2]}$.
    Multiply: #13 × #15; extract the square root.

$$\sqrt{(569)(805)} = \sqrt{458,045} = 676.8$$

17. Obtain: $r = \dfrac{N\Sigma XY - (\Sigma X)(\Sigma Y)}{\sqrt{[N\Sigma X^2 - (\Sigma X)^2][N\Sigma Y^2 - (\Sigma Y)^2]}}$.
    Divide: #11 by #16.

$$\frac{605}{676.8} = +.894$$

E. SIGNIFICANCE TEST FOR r

Enter Appendix 9 with the value of $df = N - 2$ to determine whether r is significant. If $r_{\text{Obtained}} > r_{\text{Tabled}}$, reject the null hypothesis that the population correlation equals 0.

An r of .632 is required with 8 df for significance for $\alpha = .05$.
.894 > .632.
Therefore the null hypothesis that the population correlation
equals 0 may be rejected.

F. INTERPRETATION

There is a significant, strong positive relationship between scores
on the SVT and the NAT. Trainees who score high on the SVT
prior to training tend to score high on the NAT after training,
whereas trainees who score low on the SVT tend to score low on
the NAT. Nearly 80% of the variance in NAT scores can be accounted for by the SVT scores ($r^2 = .799$).

Description:

Two variables

| SPEARMAN RHO (ρ) |

For two ordered variables or for
    score and ordered data
Each subject must be ranked on
    each variable

Follow through the steps below.

● *Verbal equivalents for all notation are presented in the* Computations *section below.*

● *If you are experienced with computations for the Spearman rho, go
directly to the computing formula below. If you require step-by-
step guidance, begin with Section B,* The Example.

A. COMPUTING FORMULA

$$\rho = 1 - \left\{ \frac{6\Sigma d_i^2}{N^3 - N} \right\}$$

where N = the number of paired rankings, and
$d_i = R_x - R_y$: each subject's rank on the X variable minus his rank on the Y variable.

NOTE: If subjects are ranked on one variable and have numerical scores on the other, convert the numerical scores to ranks. For interpretation purposes, keep in mind whether high ranks are associated with high scores or with low scores.

B. THE EXAMPLE

1. A physician was interested in determining whether alertness in the early morning is related to alertness in the late evening. He selected eight subjects and asked another physician, who was unaware of the nature of the experiment, to rank the alertness of the subjects. This physician ranked the subjects at 5:00 AM; the most alert subject was ranked 1. At 12:00 midnight, the second physician again ranked the alertness of the subjects.
2. *Null Hypothesis*
   a. *General*. The observations are drawn from a population in which the correlation between the variables is zero.
   b. *Specific*. In the population being sampled, the correlation between rated early morning alertness and late evening alertness is zero.
3. *Steps through the Branching Program for This Example*
   a. *Statistical technique needed:* **correlation coefficient**
   b. *Kind of data:* **variable 1, ordered data:** each subject was ranked on alertness in the early morning. **Variable 2, ordered data:** each subject was ranked on alertness in the late evening.

C. RECORDING THE DATA

Each subject has two ranks, one on each variable. Record the data in a table (see Table 12.2) having the following five headings:
1. *Subjects*. List each subject by number or by some other identifying rubric.
2. R_x. List the rank for each subject on one of the variables.
3. R_y. List the corresponding rank for each subject on the second variable.
4. d_i. Subtract: R_y from R_x.
5. d_i^2. Square each d_i.

Table 12.2

Subject Number	5:00 AM R_x	12:00 PM R_y	d_i	d_i^2
1	5	3	2	4
2	2	7	−5	25
3	4	5	−1	1
4	8	1	7	49
5	6	4	2	4
6	1	8	−7	49
7	3	6	−3	9
8	7	2	5	25
			Total	166

D. COMPUTATIONS

1. Obtain N.
   Count the number of paired ranks.

$$\boxed{N = 8}$$

2. Obtain Σd_i^2.
   Sum the d_i^2 column.

$$\boxed{4 + 25 + \cdots + 25 = 166}$$

3. Obtain $6\Sigma d_i^2$.

$$\boxed{6(166) = 996}$$

4. Obtain $N^3 - N$.

$$\boxed{512 - 8 = 504}$$

5. Obtain $\dfrac{6\Sigma d_i^2}{N^3 - N}$.
   Divide: #3 by #4.

$$\boxed{\frac{996}{504} = 1.976}$$

6. Obtain: $\rho = 1 - \left\{ \dfrac{6\Sigma d_i^2}{N^3 - N} \right\}$.

Subtract: $1.00 - \#5$.

$$\boxed{1 - 1.976 = -.976}$$

E. SIGNIFICANCE TEST FOR RHO

Enter Appendix 10 with the value of N, the number of subjects, to determine whether ρ is significant. If the ρ_{Obtained} is greater than the ρ_{Tabled}, reject the null hypothesis that the population correlation equals 0.

A ρ of .738 is required with $N = 8$ for significance for $\alpha = .05$.
$.976 > .738$.
Therefore the null hypothesis that the population correlation equals 0 may be rejected.

F. INTERPRETATION

There is a high, almost perfect, negative correlation between alertness in the early morning and alertness in the late evening. The more alert the subjects were in the morning, the less alert they were at night; the less alert they were in the morning, the more alert they were at night.

Description:

$\boxed{C}$
Two variables
Both category (frequency) data
Each subject is classified in one category
on each variable

Follow through the steps below.

● C *is easily computed after a* χ^2 *has been obtained for the data. If
you have already computed a* χ^2, *go directly to the computing formula below. If you have not computed a* χ^2, *begin with Section B,*
The Example.

A. COMPUTING FORMULA

$$C = \sqrt{\frac{\chi^2}{\chi^2 + N}}$$

where χ^2 = the numerical value of χ^2, obtained by performing a
χ^2 test on the data, and
N = the total number of subjects.

B. THE EXAMPLE

1. We shall use the example that was used to illustrate a 2 × 2 χ^2
in Chapter 5. That example dealt with the number of men and
women who passed or failed a particular test.
2. *Null Hypothesis*
   a. *General.* The observations are drawn from a population in
   which the relationship between the variables is zero.
   b. *Specific.* In the population being sampled, there is no relationship between gender and passing or failing a particular test.
3. *Steps through the Branching Program for This Example*
   a. *Statistical technique needed:* **correlation coefficient**
   b. *Kind of data:* **variable 1, category (frequency) data**: each
   subject is classified as a man or a woman. **Variable 2, category
   (frequency) data**: each subject is classified as passing or failing the test.

C. RECORDING THE DATA

1. Record the data in an *a* × *b* contingency table. Go to Chapter
5, p. 74, for details on computing χ^2.

	Fail	Succeed	
Women	13	26	39
Men	26	10	36
	39	36	75

D. COMPUTATIONS

1. Obtain χ^2.
   Perform appropriate calculations as detailed in Chapter 5.

$$\chi^2 = \frac{75(|676 - 130| - 75/2)^2}{(39)(36)(39)(36)} = 9.84$$

2. Obtain $C = \sqrt{\dfrac{\chi^2}{\chi^2 + N}}$.

   Divide the value of χ^2 by the value of $(\chi^2 + N)$. Obtain the square root of the value.

$$C = \sqrt{\frac{9.84}{9.84 + 75}} = \sqrt{\frac{9.84}{84.84}} = \sqrt{.11598} = .341$$

E. SIGNIFICANCE TEST FOR *C*

The χ^2 used in the computation of C is the statistic by which the significance of C is tested. Evaluate χ^2 with the appropriate df in Appendix 1 (see Chapter 5). A significant χ^2 indicates that C is significant. If C is significant, reject the null hypothesis that the population relationship equals 0.

With 1 df, χ^2 of 3.84 is needed for significance for $\alpha = .05$. 9.84 > 3.84.
Since χ^2 is significant at .05, C is significant at .05.
The null hypothesis that the population correlation equals 0 may be rejected.

F. INTERPRETATION

There is a moderately low but significant relationship between gender of the subjects and success or failure on the test: women were more likely to succeed and men were more likely to fail.

Description:

ETA (η)

Two variables

Score data and category (frequency) data

Each subject has a numerical score on one variable and is classified in one category on the other variable

Follow through the steps below.

- η *is most easily obtained after a one-way analysis of variance has been performed on the data. If you have already computed an F ratio and are experienced with computations for η, go directly to the computing formula below. If you require step-by-step guidance, begin with Section B,* The Example.

A. COMPUTING FORMULA[7]

$$\eta = \sqrt{\frac{(df_A)(F)}{(df_A)(F) + df_{\text{Error}}}}$$

where F = the numerical value of the F ratio,

     df_A = the degrees of freedom for A, the treatment, and

     df_{Error} = the degrees of freedom for error.

B. THE EXAMPLE

1. We shall use the example from Chapter 7 that was used to illustrate a 1 X 3 analysis of variance. That example dealt with the number of feet that high school boys were able to put the shot under three incentive conditions.
2. *Null Hypothesis*
   a. *General.* The observations are drawn from a population in which the relationship between the variables is zero.
   b. *Specific.* In the population being sampled, there is no relationship between incentive conditions and the distance that the shot was put.
3. *Steps through the Branching Program for This Example*
   a. *Statistical technique needed:* **correlation coefficient**

---

[7] The careful reader will note that this formula is quite different from the one given in Chapter 11, p. 334, which reads $\eta^2 = SS_x/SS_{\text{Total}}$.   The formula in Chapter 11 is more general (it applies to one-way, two-way, etc. ANOVA), whereas the present formula applies only to the one-way ANOVA.

b. *Kind of data:* **variable 1, category (frequency) data:** each subject is classified as falling into one of the three incentive conditions. **Variable 2, score data:** each subject has a score representing how far he put the shot.

C. RECORDING THE DATA

1. Record the data in a 1 $\times$ a matrix. See Chapter 7, p. 138, for complete instructions for computing a one-way analysis of variance. Summarize the analysis in a summary table (see Table 12.3).

Table 12.3. Summary Table: One-Way ANOVA (1 $\times$ a)

Source	df	SS	MS	F
Incentive (A)	2	251.72	125.86	10.89
Error	15	173.39	11.56	
Total	17	425.11		

D. COMPUTATIONS

1. Obtain F.
Perform appropriate calculations as detailed in Chapter 7.

$$F = 10.89$$

2. Determine:
a. df_A, the degrees of freedom for the treatment, and
b. df_{Error}, the degrees of freedom for the error term in the analysis of variance.

$$df_A = 2, \ df_{Error} = 15$$

3. Obtain $(df_A)(F)$.
Multiply: #2a $\times$ #1.

$$2(10.89) = 21.78$$

4. Obtain $df_A(F) + df_{Error}$.
Add: #3 + #2b.

$$\boxed{21.78 + 15 = 36.78}$$

5. Obtain $\eta = \sqrt{\dfrac{(df_A)(F)}{(df_A)(F) + df_{\text{Error}}}}$.

Divide #3 by #4, then extract the square root.

$$\boxed{\sqrt{\dfrac{21.78}{36.78}} = \sqrt{.59216} = .769}$$

E. SIGNIFICANCE TEST FOR η

The F used in the computation of η is the statistic by which the significance of η is tested. Evaluate F with the appropriate df in Appendix 3. A significant F indicates that η is significant. If η is significant, reject the null hypothesis that the population correlation equals 0.

> With $df = 1, 15$, F of 4.54 is needed for significance for $\alpha = .05$. 10.89 > 4.54.
> Since F is significant at .05, η is significant at .05.
> The null hypothesis that the population correlation equals 0 may be rejected.

F. INTERPRETATION

There is a relatively high significant relationship between the incentive conditions and the distance that the shot was put. The incentive conditions accounted for almost 60% ($\eta^2 = .592$) of the variance in the shot-put scores.

Description:

ETA (η)	Two variables
	Ordered data and category (frequency) data
	Each subject has a rank on one variable and is classified in one category on the other variable

Follow through the steps below.

● η *is most easily calculated after a rank-sums test (if there are two categories) or a Kruskal-Wallis test (if there are more than two categories) has been performed on the data. If you have already computed a Z (rank-sums) or an H (Kruskal-Wallis), go directly to the computing formula below. If you have not computed a Z or an H, begin with Section B, The Examples.*

A. COMPUTING FORMULAS

1. If there are only two categories:

$$\eta = \sqrt{\frac{Z^2}{N-1}}$$

where Z = the numerical value of Z from the rank-sums test, and
N — the number of scores.

2. If there are more than two categories:

$$\eta = \sqrt{\frac{H}{N-1}}$$

where H = the numerical value of H from the Kruskal-Wallis test, and
N = the number of scores.

B. THE EXAMPLES

1. We shall use the example given to illustrate the rank-sums test in Chapter 6, which dealt with the relationship between general appearance and vitality and type of diet in a group of 21 Labrador puppies.

2. We shall use the example given to illustrate the Kruskal-Wallis test in Chapter 6, which dealt with the relationship between marital status and general adjustment in a sample of 24 women.

3. *Null Hypothesis*
   a. *General.* The observations are drawn from a population in which the relationship between the variables is zero.
   b. *Specific. Example 1.* In the population being sampled, there is no relationship between dogs' appearance and vitality, and the type of diet they were fed. *Example 2.* In the population being sampled, there is no relationship between marital status and general adjustment.

4. *Steps through the Branching Program for These Examples*
   Example 1:
   a. *Statistical technique needed:* **correlation coefficient**
   b. *Kind of data:* **variable 1, category (frequency) data:** each dog is classified as being in one of the two dietary groups. **Variable 2, ordered data:** each dog is ranked on overall appearance and vitality.
   Example 2:
   a. *Statistical technique needed:* **correlation coefficient**
   b. *Kind of data:* **variable 1, category (frequency) data:** each woman is classified as being in one of three marital-status groups. **Variable 2, ordered data:** each woman is ranked on general adjustment.

C. RECORDING THE DATA

1. Record the data in appropriate tabular form so that the rank-sums test (two categories) (see Table 12.4) or the Kruskal-Wallis test (more than two categories) (see Table 12.5) can be performed. Go to Chapter 6 for complete instructions for computing a rank-sums or Kruskal-Wallis test.

Table 12.4. Data Table: Rank-Sums Test
Example 1

Rank	Group	DS Group	NF Group
1	DS	1	
2	DS	2	
3.5	DS	3.5	
3.5	DS	3.5	
5	NF		5
6	DS	6	
7	DS	7	
8	DS	8	
9	NF		9
10	DS	10	
11.5	DS	11.5	
11.5	DS	11.5	
13	NF		13
14	NF		14
15	NF		15
16	NF		16
17	NF		17
18	DS	18	
19	NF		19
20	NF		20
21	NF		21
		$T_1 = 82$	$T_2 = 149$
		$N_1 = 11$	$N_2 = 10$

Table 12.5. Data Table: Kruskal-Wallis Test
Example 2

Married		Divorced		Single	
Score	Rank	Score	Rank	Score	Rank
18	3.5	12	1	18	3.5
28	8	16	2	21	5
32	9	37	11	26	6.5
46	13.5	40	12	26	6.5
52	16.5	46	13.5	33	10
62	20	52	16.5	51	15
63	21.5	61	19	53	18
		63	21.5	68	23
				70	24
$T_1 = 92.0$		$T_2 = 96.5$		$T_3 = 111.5$	
$n_1 = 7$		$n_2 = 8$		$n_3 = 9$	

D. COMPUTATIONS

1. Obtain Z (two categories) or H (more than two categories). Perform appropriate calculations as detailed in Chapter 6.

> Example 1:
> $Z = -2.75$ (From example, Chapter 6, p. 106)

> Example 2:
> $H = .09$ (From example, Chapter 6, p. 111)

2. Obtain η:
    a. For the rank-sums test,

$$\eta = \sqrt{\frac{Z^2}{N-1}}.$$

Square Z, divide by $N - 1$, and extract the square root.

    b. For the Kruskal-Wallis test,

$$\eta = \sqrt{\frac{H}{N-1}}.$$

Divide H by $N - 1$. Extract the square root.

> Example 1:
> $$\eta = \sqrt{\frac{Z^2}{N-1}} = \sqrt{\frac{-2.75^2}{21-1}} = \sqrt{\frac{7.5625}{20}} = \sqrt{.378} = .615$$

> Example 2:
> $$\eta = \sqrt{\frac{H}{N-1}} = \sqrt{\frac{.09}{24-1}} = \sqrt{\frac{.09}{23}} = \sqrt{.003913} = .063$$

E. SIGNIFICANCE TEST FOR η

The Z or H used in the computation of η is the statistic by which the significance of η is tested. Evaluate Z with normal curve ($Z = \pm 1.96$ for $\alpha = .05$) or H with table of χ^2, Appendix 1. If Z or H is significant, η is significant. If η is significant, reject the null hypothesis that the population correlation equals 0.

> Example 1:
> Z of ±1.96 is needed for significance for $\alpha = .05$.
> $-2.75 < -1.96$.
> Since Z is significant for $\alpha = .05$, η is significant.
> The null hypothesis that the population correlation
> equals 0 may be rejected.

> Example 2:
> With 2 *df*, an H of 6.0 is needed for significance for $\alpha = .05$.
> $.09 < 6.0$.
> Since H is not significant for $\alpha = .05$, η is not significant.
> The null hypothesis that the population correlation equals 0
> cannot be rejected.

F. INTERPRETATION

Example 1. There is a moderate but significant relationship between the dogs' appearance and vitality and the type of diet they were fed. The diet accounted for 37.8% of the variance in the rankings of vitality and general appearance ($\eta^2 = .378$).

Example 2. Marital status is not related to general adjustment in the population from which this sample was drawn.

APPENDIX 1

Values of chi square (χ^2) for $\alpha = .05$ and $.01$

df	$.05$	$.01$
1	3.84146	6.63490
2	5.99147	9.21034
3	7.81473	11.3449
4	9.48773	13.2767
5	11.0705	15.0863
6	12.5916	16.8119
7	14.0671	18.4753
8	15.5073	20.0902
9	16.9190	21.6660
10	18.3070	23.2093
11	19.6751	24.7250
12	21.0261	26.2170
13	22.3621	27.6883
14	23.6848	29.1413
15	24.9958	30.5779
16	26.2962	31.9999
17	27.5871	33.4087
18	28.8693	34.8053
19	30.1435	36.1908
20	31.4104	37.5662
21	32.6705	38.9321
22	33.9244	40.2894
23	35.1725	41.6384
24	36.4151	42.9798
25	37.6525	44.3141
26	38.8852	45.6417
27	40.1133	46.9630
28	41.3372	48.2782
29	42.5569	49.5879
30	43.7729	50.8922
40	55.7585	63.6907
50	67.5048	76.1539
60	79.0819	88.3794
70	90.5312	100.425
80	101.879	112.329
90	113.145	124.116
100	124.342	135.807

APPENDIX 2

Critical values of R for the sign test for $\alpha = .01$ and $.05$

N	$.01$	$.05$	N	$.01$	$.05$
1			46	13	15
2			47	14	16
3			48	14	16
4			49	15	17
5			50	15	17
6		0	51	15	18
7		0	52	16	18
8	0	0	53	16	18
9	0	1	54	17	19
10	0	1	55	17	19
11	0	1	56	17	20
12	1	2	57	18	20
13	1	2	58	18	21
14	1	2	59	19	21
15	2	3	60	19	21
16	2	3	61	20	22
17	2	4	62	20	22
18	3	4	63	20	23
19	3	4	64	21	23
20	3	5	65	21	24
21	4	5	66	22	24
22	4	5	67	22	25
23	4	6	68	22	25
24	5	6	69	23	25
25	5	7	70	23	26
26	6	7	71	24	26
27	6	7	72	24	27
28	6	8	73	25	27
29	7	8	74	25	28
30	7	9	75	25	28
31	7	9	76	26	28
32	8	9	77	26	29
33	8	10	78	27	29
34	9	10	79	27	30
35	9	11	80	28	30
36	9	11	81	28	31
37	10	12	82	28	31
38	10	12	83	29	32
39	11	12	84	29	32
40	11	13	85	30	32
41	11	13	86	30	33
42	12	14	87	31	33
43	12	14	88	31	34
44	13	15	89	31	34
45	13	15	90	32	35

From W. J. Dixon and F. J. Massey, Jr., *Introduction to Statistical Analysis*, 3rd ed., McGraw-Hill, 1969, p. 509. Reproduced by permission.

APPENDIX 3

Values of F for $\alpha = .05$

								df_1 = Degrees of Freedom for Numerator											
Degrees of Freedom for Denominator df_2	1	2	3	4	5	6	7	8	9	10	12	15	20	24	30	40	60	120	∞
1	161.4	199.5	215.7	224.6	230.2	234.0	236.8	238.9	240.5	241.9	243.9	245.9	248.0	249.1	250.1	251.1	252.2	253.3	254.3
2	18.51	19.00	19.16	19.25	19.30	19.33	19.35	19.37	19.38	19.40	19.41	19.43	19.45	19.45	19.46	19.47	19.48	19.49	19.50
3	10.13	9.55	9.28	9.12	9.01	8.94	8.89	8.85	8.81	8.79	8.74	8.70	8.66	8.64	8.62	8.59	8.57	8.55	8.53
4	7.71	6.94	6.59	6.39	6.26	6.16	6.09	6.04	6.00	5.96	5.91	5.86	5.80	5.77	5.75	5.72	5.69	5.65	5.63
5	6.61	5.79	5.41	5.19	5.05	4.95	4.88	4.82	4.77	4.74	4.68	4.62	4.56	4.53	4.50	4.46	4.43	4.40	4.36
6	5.99	5.14	4.76	4.53	4.39	4.28	4.21	4.15	4.10	4.06	4.00	3.94	3.87	3.84	3.81	3.77	3.74	3.70	3.67
7	5.59	4.74	4.35	4.12	3.97	3.87	3.79	3.73	3.68	3.64	3.57	3.51	3.44	3.41	3.38	3.34	3.30	3.27	3.23
8	5.32	4.46	4.07	3.84	3.69	3.58	3.50	3.44	3.39	3.35	3.28	3.22	3.15	3.12	3.08	3.04	3.01	2.97	2.93
9	5.12	4.26	3.86	3.63	3.48	3.37	3.29	3.23	3.18	3.14	3.07	3.01	2.94	2.90	2.86	2.83	2.79	2.75	2.71
10	4.96	4.10	3.71	3.48	3.33	3.22	3.14	3.07	3.02	2.98	2.91	2.85	2.77	2.74	2.70	2.66	2.62	2.58	2.54
11	4.84	3.98	3.59	3.36	3.20	3.09	3.01	2.95	2.90	2.85	2.79	2.72	2.65	2.61	2.57	2.53	2.49	2.45	2.40
12	4.75	3.89	3.49	3.26	3.11	3.00	2.91	2.85	2.80	2.75	2.69	2.62	2.54	2.51	2.47	2.43	2.38	2.34	2.30
13	4.67	3.81	3.41	3.18	3.03	2.92	2.83	2.77	2.71	2.67	2.60	2.53	2.46	2.42	2.38	2.34	2.30	2.25	2.21
14	4.60	3.74	3.34	3.11	2.96	2.85	2.76	2.70	2.65	2.60	2.53	2.46	2.39	2.35	2.31	2.27	2.22	2.18	2.13
15	4.54	3.68	3.29	3.06	2.90	2.79	2.71	2.64	2.59	2.54	2.48	2.40	2.33	2.29	2.25	2.20	2.16	2.11	2.07
16	4.49	3.63	3.24	3.01	2.85	2.74	2.66	2.59	2.54	2.49	2.42	2.35	2.28	2.24	2.19	2.15	2.11	2.06	2.01
17	4.45	3.59	3.20	2.96	2.81	2.70	2.61	2.55	2.49	2.45	2.38	2.31	2.23	2.19	2.15	2.10	2.06	2.01	1.96
18	4.41	3.55	3.16	2.93	2.77	2.66	2.58	2.51	2.46	2.41	2.34	2.27	2.19	2.15	2.11	2.06	2.02	1.97	1.92
19	4.38	3.52	3.13	2.90	2.74	2.63	2.54	2.48	2.42	2.38	2.31	2.23	2.16	2.11	2.07	2.03	1.98	1.93	1.88
20	4.35	3.49	3.10	2.87	2.71	2.60	2.51	2.45	2.39	2.35	2.28	2.20	2.12	2.08	2.04	1.99	1.95	1.90	1.84
21	4.32	3.47	3.07	2.84	2.68	2.57	2.49	2.42	2.37	2.32	2.25	2.18	2.10	2.05	2.01	1.96	1.92	1.87	1.81
22	4.30	3.44	3.05	2.82	2.66	2.55	2.46	2.40	2.34	2.30	2.23	2.15	2.07	2.03	1.98	1.94	1.89	1.84	1.78
23	4.28	3.42	3.03	2.80	2.64	2.53	2.44	2.37	2.32	2.27	2.20	2.13	2.05	2.01	1.96	1.91	1.86	1.81	1.76
24	4.26	3.40	3.01	2.78	2.62	2.51	2.42	2.36	2.30	2.25	2.18	2.11	2.03	1.98	1.94	1.89	1.84	1.79	1.73
25	4.24	3.39	2.99	2.76	2.60	2.49	2.40	2.34	2.28	2.24	2.16	2.09	2.01	1.96	1.92	1.87	1.82	1.77	1.71
26	4.23	3.37	2.98	2.74	2.59	2.47	2.39	2.32	2.27	2.22	2.15	2.07	1.99	1.95	1.90	1.85	1.80	1.75	1.69
27	4.21	3.35	2.96	2.73	2.57	2.46	2.37	2.31	2.25	2.20	2.13	2.06	1.97	1.93	1.88	1.84	1.79	1.73	1.67
28	4.20	3.34	2.95	2.71	2.56	2.45	2.36	2.29	2.24	2.19	2.12	2.04	1.96	1.91	1.87	1.82	1.77	1.71	1.65
29	4.18	3.33	2.93	2.70	2.55	2.43	2.35	2.28	2.22	2.18	2.10	2.03	1.94	1.90	1.85	1.81	1.75	1.70	1.64
30	4.17	3.32	2.92	2.69	2.53	2.42	2.33	2.27	2.21	2.16	2.09	2.01	1.93	1.89	1.84	1.79	1.74	1.68	1.62
40	4.08	3.23	2.84	2.61	2.45	2.34	2.25	2.18	2.12	2.08	2.00	1.92	1.84	1.79	1.74	1.69	1.64	1.58	1.51
60	4.00	3.15	2.76	2.53	2.37	2.25	2.17	2.10	2.04	1.99	1.92	1.84	1.75	1.70	1.65	1.59	1.53	1.47	1.39
120	3.92	3.07	2.68	2.45	2.29	2.17	2.09	2.02	1.96	1.91	1.83	1.75	1.66	1.61	1.55	1.50	1.43	1.35	1.25
∞	3.84	3.00	2.60	2.37	2.21	2.10	2.01	1.94	1.88	1.83	1.75	1.67	1.57	1.52	1.46	1.39	1.32	1.22	1.00

APPENDIX 3 (continued)

Values of F for $\alpha = .01$

$df_1 = $ Degrees of Freedom for Numerator

Degrees of Freedom for Denominator df_2	1	2	3	4	5	6	7	8	9	10	12	15	20	24	30	40	60	120	∞
1	4052	4999.5	5403	5625	5764	5859	5928	5981	6022	6056	6106	6157	6209	6235	6261	6287	6313	6339	6366
2	98.50	99.00	99.17	99.25	99.30	99.33	99.36	99.37	99.39	99.40	99.42	99.43	99.45	99.46	99.47	99.47	99.48	99.49	99.50
3	34.12	30.82	29.46	28.71	28.24	27.91	27.67	27.49	27.35	27.23	27.05	26.87	26.69	26.60	26.50	26.41	26.32	26.22	26.13
4	21.20	18.00	16.69	15.98	15.52	15.21	14.98	14.80	14.66	14.55	14.37	14.20	14.02	13.93	13.84	13.75	13.65	13.56	13.46
5	16.26	13.27	12.06	11.39	10.97	10.67	10.46	10.29	10.16	10.05	9.89	9.72	9.55	9.47	9.38	9.29	9.20	9.11	9.02
6	13.75	10.92	9.78	9.15	8.75	8.47	8.26	8.10	7.98	7.87	7.72	7.56	7.40	7.31	7.23	7.14	7.06	6.97	6.88
7	12.25	9.55	8.45	7.85	7.46	7.19	6.99	6.84	6.72	6.62	6.47	6.31	6.16	6.07	5.99	5.91	5.82	5.74	5.65
8	11.26	8.65	7.59	7.01	6.63	6.37	6.18	6.03	5.91	5.81	5.67	5.52	5.36	5.28	5.20	5.12	5.03	4.95	4.86
9	10.56	8.02	6.99	6.42	6.06	5.80	5.61	5.47	5.35	5.26	5.11	4.96	4.81	4.73	4.65	4.57	4.48	4.40	4.31
10	10.04	7.56	6.55	5.99	5.64	5.39	5.20	5.06	4.94	4.85	4.71	4.56	4.41	4.33	4.25	4.17	4.08	4.00	3.91
11	9.65	7.21	6.22	5.67	5.32	5.07	4.89	4.74	4.63	4.54	4.40	4.25	4.10	4.02	3.94	3.86	3.78	3.69	3.60
12	9.33	6.93	5.95	5.41	5.06	4.82	4.64	4.50	4.39	4.30	4.16	4.01	3.86	3.78	3.70	3.62	3.54	3.45	3.36
13	9.07	6.70	5.74	5.21	4.86	4.62	4.44	4.30	4.19	4.10	3.96	3.82	3.66	3.59	3.51	3.43	3.34	3.25	3.17
14	8.86	6.51	5.56	5.04	4.69	4.46	4.28	4.14	4.03	3.94	3.80	3.66	3.51	3.43	3.35	3.27	3.18	3.09	3.00
15	8.68	6.36	5.42	4.89	4.56	4.32	4.14	4.00	3.89	3.80	3.67	3.52	3.37	3.29	3.21	3.13	3.05	2.96	2.87
16	8.53	6.23	5.29	4.77	4.44	4.20	4.03	3.89	3.78	3.69	3.55	3.41	3.26	3.18	3.10	3.02	2.93	2.84	2.75
17	8.40	6.11	5.18	4.67	4.34	4.10	3.93	3.79	3.68	3.59	3.46	3.31	3.16	3.08	3.00	2.92	2.83	2.75	2.65
18	8.29	6.01	5.09	4.58	4.25	4.01	3.84	3.71	3.60	3.51	3.37	3.23	3.08	3.00	2.92	2.84	2.75	2.66	2.57
19	8.18	5.93	5.01	4.50	4.17	3.94	3.77	3.63	3.52	3.43	3.30	3.15	3.00	2.92	2.84	2.76	2.67	2.58	2.49
20	8.10	5.85	4.94	4.43	4.10	3.87	3.70	3.56	3.46	3.37	3.23	3.09	2.94	2.86	2.78	2.69	2.61	2.52	2.42
21	8.02	5.78	4.87	4.37	4.04	3.81	3.64	3.51	3.40	3.31	3.17	3.03	2.88	2.80	2.72	2.64	2.55	2.46	2.36
22	7.95	5.72	4.82	4.31	3.99	3.76	3.59	3.45	3.35	3.26	3.12	2.98	2.83	2.75	2.67	2.58	2.50	2.40	2.31
23	7.88	5.66	4.76	4.26	3.94	3.71	3.54	3.41	3.30	3.21	3.07	2.93	2.78	2.70	2.62	2.54	2.45	2.35	2.26
24	7.82	5.61	4.72	4.22	3.90	3.67	3.50	3.36	3.26	3.17	3.03	2.89	2.74	2.66	2.58	2.49	2.40	2.31	2.21
25	7.77	5.57	4.68	4.18	3.85	3.63	3.46	3.32	3.22	3.13	2.99	2.85	2.70	2.62	2.54	2.45	2.36	2.27	2.17
26	7.72	5.53	4.64	4.14	3.82	3.59	3.42	3.29	3.18	3.09	2.96	2.81	2.66	2.58	2.50	2.42	2.33	2.23	2.13
27	7.68	5.49	4.60	4.11	3.78	3.56	3.39	3.26	3.15	3.06	2.93	2.78	2.63	2.55	2.47	2.38	2.29	2.20	2.10
28	7.64	5.45	4.57	4.07	3.75	3.53	3.36	3.23	3.12	3.03	2.90	2.75	2.60	2.52	2.44	2.35	2.26	2.17	2.06
29	7.60	5.42	4.54	4.04	3.73	3.50	3.33	3.20	3.09	3.00	2.87	2.73	2.57	2.49	2.41	2.33	2.23	2.14	2.03
30	7.56	5.39	4.51	4.02	3.70	3.47	3.30	3.17	3.07	2.98	2.84	2.70	2.55	2.47	2.39	2.30	2.21	2.11	2.01
40	7.31	5.18	4.31	3.83	3.51	3.29	3.12	2.99	2.89	2.80	2.66	2.52	2.37	2.29	2.20	2.11	2.02	1.92	1.80
60	7.08	4.98	4.13	3.65	3.34	3.12	2.95	2.82	2.72	2.63	2.50	2.35	2.20	2.12	2.03	1.94	1.84	1.73	1.60
120	6.85	4.79	3.95	3.48	3.17	2.96	2.79	2.66	2.56	2.47	2.34	2.19	2.03	1.95	1.86	1.76	1.66	1.53	1.38
∞	6.63	4.61	3.78	3.32	3.02	2.80	2.64	2.51	2.41	2.32	2.18	2.04	1.88	1.79	1.70	1.59	1.47	1.32	1.00

Abridged from E. S. Pearson and H. O. Hartley, *Biometrika Tables for Statisticians*, Vol. 1, 3rd ed., Cambridge University Press. Reproduced by permission of the authors and the *Biometrika* trustees.

APPENDIX 4

Critical values of the t distribution for $\alpha = .05$ and $.01$

Degrees of Freedom	.05	.01
1	12.706	63.657
2	4.303	9.925
3	3.182	5.841
4	2.776	4.604
5	2.571	4.032
6	2.447	3.707
7	2.365	3.499
8	2.306	3.355
9	2.262	3.250
10	2.228	3.169
11	2.201	3.106
12	2.179	3.055
13	2.160	3.012
14	2.145	2.977
15	2.131	2.947
16	2.120	2.921
17	2.110	2.898
18	2.101	2.878
19	2.093	2.861
20	2.086	2.845
21	2.080	2.831
22	2.074	2.819
23	2.069	2.807
24	2.064	2.797
25	2.060	2.787
26	2.056	2.779
27	2.052	2.771
28	2.048	2.763
29	2.045	2.756
30	2.042	2.750
40	2.021	2.704
60	2.000	2.660
120	1.980	2.617
∞	1.960	2.576

APPENDIX 5

Extended values of χ^2 for use with Ryan's procedure

a	7	6	5	$d - 1$ 4	3	2	1
8	9.80	9.55	9.18	8.76	8.18	7.51	6.25
7		9.24	8.94	8.53	8.01	7.24	6.00
6			8.64	8.24	7.67	6.92	5.76
5				7.90	7.40	6.66	5.43
4					6.97	6.25	5.02
3						5.76	4.54

APPENDIX 6

Extended table of Z for use with Ryan's procedure for pairwise comparisons using the rank-sums test to make multiple comparisons with the Kruskal-Wallis test

a	7	6	5	$d - 1$ 4	3	2	1
8	3.13	3.09	3.03	2.96	2.86	2.74	2.50
7		3.04	2.99	2.92	2.83	2.69	2.45
6			2.94	2.87	2.77	2.63	2.40
5				2.81	2.72	2.58	2.33
4					2.64	2.50	2.24
3						2.40	2.13

APPENDIX 7

Distribution of the studentized range statistic, q_k or q_d

df for MS_{Error}	α	d = Number of Ordered Means Spanned by the Comparison or k = Number of Means Being Compared													
		2	3	4	5	6	7	8	9	10	11	12	13	14	15
1	.05	18.0	27.0	32.8	37.1	40.4	43.1	45.4	47.4	49.1	50.6	52.0	53.2	54.3	55.4
	.01	90.0	135	164	186	202	216	227	237	246	253	260	266	272	277
2	.05	6.09	8.3	9.8	10.9	11.7	12.4	13.0	13.5	14.0	14.4	14.7	15.1	15.4	15.7
	.01	14.0	19.0	22.3	24.7	26.6	28.2	29.5	30.7	31.7	32.6	33.4	34.1	34.8	35.4
3	.05	4.50	5.91	6.82	7.50	8.04	8.48	8.85	9.18	9.46	9.72	9.95	10.2	10.4	10.5
	.01	8.26	10.6	12.2	13.3	14.2	15.0	15.6	16.2	16.7	17.1	17.5	17.9	18.2	18.5
4	.05	3.93	5.04	5.76	6.29	6.71	7.05	7.35	7.60	7.83	8.03	8.21	8.37	8.52	8.66
	.01	6.51	8.12	9.17	9.96	10.6	11.1	11.5	11.9	12.3	12.6	12.8	13.1	13.3	13.5
5	.05	3.64	4.60	5.22	5.67	6.03	6.33	6.58	6.80	6.99	7.17	7.32	7.47	7.60	7.72
	.01	5.70	6.97	7.80	8.42	8.91	9.32	9.67	9.97	10.2	10.5	10.7	10.9	11.1	11.2
6	.05	3.46	4.34	4.90	5.31	5.63	5.89	6.12	6.32	6.49	6.65	6.79	6.92	7.03	7.14
	.01	5.24	6.33	7.03	7.56	7.97	8.32	8.61	8.87	9.10	9.30	9.49	9.65	9.81	9.95
7	.05	3.34	4.16	4.69	5.06	5.36	5.61	5.82	6.00	6.16	6.30	6.43	6.55	6.66	6.76
	.01	4.95	5.92	6.54	7.01	7.37	7.68	7.94	8.17	8.37	8.55	8.71	8.86	9.00	9.12
8	.05	3.26	4.04	4.53	4.89	5.17	5.40	5.60	5.77	5.92	6.05	6.18	6.29	6.39	6.48
	.01	4.74	5.63	6.20	6.63	6.96	7.24	7.47	7.68	7.87	8.03	8.18	8.31	8.44	8.55
9	.05	3.20	3.95	4.42	4.76	5.02	5.24	5.43	5.60	5.74	5.87	5.98	6.09	6.19	6.28
	.01	4.60	5.43	5.96	6.35	6.66	6.91	7.13	7.32	7.49	7.65	7.78	7.91	8.03	8.13
10	.05	3.15	3.88	4.33	4.65	4.91	5.12	5.30	5.46	5.60	5.72	5.83	5.93	6.03	6.11
	.01	4.48	5.27	5.77	6.14	6.43	6.67	6.87	7.05	7.21	7.36	7.48	7.60	7.71	7.81
11	.05	3.11	3.82	4.26	4.57	4.82	5.03	5.20	5.35	5.49	5.61	5.71	5.81	5.90	5.99
	.01	4.39	5.14	5.62	5.97	6.25	6.48	6.67	6.84	6.99	7.13	7.26	7.36	7.46	7.56

df for MS_Error	α	d = Number of Ordered Means Spanned by the Comparison or k = Number of Means Being Compared													
		2	3	4	5	6	7	8	9	10	11	12	13	14	15
12	.05	3.08	3.77	4.20	4.51	4.75	4.95	5.12	5.27	5.40	5.51	5.62	5.71	5.80	5.88
	.01	4.32	5.04	5.50	5.84	6.10	6.32	6.51	6.67	6.81	6.94	7.06	7.17	7.26	7.36
13	.05	3.06	3.73	4.15	4.45	4.69	4.88	5.05	5.19	5.32	5.43	5.53	5.63	5.71	5.79
	.01	4.26	4.96	5.40	5.73	5.98	6.19	6.37	6.53	6.67	6.79	6.90	7.01	7.10	7.19
14	.05	3.03	3.70	4.11	4.41	4.64	4.83	4.99	5.13	5.25	5.36	5.46	5.55	5.64	5.72
	.01	4.21	4.89	5.32	5.63	5.88	6.08	6.26	6.41	6.54	6.66	6.77	6.87	6.96	7.05
16	.05	3.00	3.65	4.05	4.33	4.56	4.74	4.90	5.03	5.15	5.26	5.35	5.44	5.52	5.59
	.01	4.13	4.78	5.19	5.49	5.72	5.92	6.08	6.22	6.35	6.46	6.56	6.66	6.74	6.82
18	.05	2.97	3.61	4.00	4.28	4.49	4.67	4.82	4.96	5.07	5.17	5.27	5.35	5.43	5.50
	.01	4.07	4.70	5.09	5.38	5.60	5.79	5.94	6.08	6.20	6.31	6.41	6.50	6.58	6.65
20	.05	2.95	3.58	3.96	4.23	4.45	4.62	4.77	4.90	5.01	5.11	5.20	5.28	5.36	5.43
	.01	4.02	4.64	5.02	5.29	5.51	5.69	5.84	5.97	6.09	6.19	6.29	6.37	6.45	6.52
24	.05	2.92	3.53	3.90	4.17	4.37	4.54	4.68	4.81	4.92	5.01	5.10	5.18	5.25	5.32
	.01	3.96	4.54	4.91	5.17	5.37	5.54	5.69	5.81	5.92	6.02	6.11	6.19	6.26	6.33
30	.05	2.89	3.49	3.84	4.10	4.30	4.46	4.60	4.72	4.83	4.92	5.00	5.08	5.15	5.21
	.01	3.89	4.45	4.80	5.05	5.24	5.40	5.54	5.56	5.76	5.85	5.93	6.01	6.08	6.14
40	.05	2.86	3.44	3.79	4.04	4.23	4.39	4.52	4.63	4.74	4.82	4.91	4.98	5.05	5.11
	.01	3.82	4.37	4.70	4.93	5.11	5.27	5.39	5.50	5.60	5.69	5.77	5.84	5.90	5.96
60	.05	2.83	3.40	3.74	3.98	4.16	4.31	4.44	4.55	4.65	4.73	4.81	4.88	4.94	5.00
	.01	3.76	4.28	4.60	4.82	4.99	5.13	5.25	5.36	5.45	5.53	5.60	5.67	5.73	5.79
120	.05	2.80	3.36	3.69	3.92	4.10	4.24	4.36	4.48	4.56	4.64	4.72	4.78	4.84	4.90
	.01	3.70	4.20	4.50	4.71	4.87	5.01	5.12	5.21	5.30	5.38	5.44	5.51	5.56	5.61
∞	.05	2.77	3.31	3.63	3.86	4.03	4.17	4.29	4.39	4.47	4.55	4.62	4.68	4.74	4.80
	.01	3.64	4.12	4.40	4.60	4.76	4.88	4.99	5.08	5.16	5.23	5.29	5.35	5.40	5.45

From B. J. Winer, *Statistical Principles in Experimental Design*, McGraw-Hill, 1962; abridged from H. L. Harter, D. S. Clemm, and E. H. Guthrie, The probability integrals of the range and of the studentized range, WADC Tech. Rep. 58–484, Vol. 2, 1959, Wright Air Development Center, Table II.2, pp. 243–281. Reproduced by permission of the authors and McGraw-Hill Book Company.

APPENDIX 8

Number of adjusted k' treatments as a function of the number of unconfounded paired comparisons in a given interaction table

Number of Unconfounded Comparisons	Number of Adjusted k' Treatments
3–4	3
5–8	4
9–12	5
13–17	6
18–24	7
25–32	8
33–40	9
41–50	10
51–60	11
61–72	12
73–84	13
85–98	14
99–112	15
113–128	16
129–144	17
145–162	18
163–180	19
181–200	20

Adapted from D. V. Cicchetti. Extensions of multiple-range tests to interaction tables in the analysis of variance: A rapid approximate solution. *Psychological Bulletin*, 1972, **77**, 405–408. Copyright 1972 by the American Psychological Association. Reproduced by permission.

APPENDIX 9

Values of r for α = .05 and .01

df	.05	.01	df	.05	.01
1	.997	1.000	24	.388	.496
2	.950	.990	25	.381	.487
3	.878	.959	26	.374	.478
4	.811	.917	27	.367	.470
5	.754	.874	28	.361	.463
6	.707	.834	29	.355	.456
7	.666	.798	30	.349	.449
8	.632	.765	35	.325	.418
9	.602	.735	40	.304	.393
10	.576	.708	45	.288	.372
11	.553	.684	50	.273	.354
12	.532	.661	60	.250	.325
13	.514	.641	70	.232	.302
14	.497	.623	80	.217	.283
15	.482	.606	90	.205	.267
16	.468	.590	100	.195	.254
17	.456	.575	125	.174	.228
18	.444	.561	150	.159	.208
19	.433	.549	200	.138	.181
20	.423	.537	300	.113	.148
21	.413	.526	400	.098	.128
22	.404	.515	500	.088	.115
23	.396	.505	1000	.062	.081

Adapted from J. T. Spence et al., *Elementary Statistics*, 2nd ed., Appleton-Century-Crofts, 1968, p. 236; abridged from Table VI of Fisher and Yates, *Statistical Tables for Biological, Agricultural, and Medical Research,* 4th ed., 1953, Oliver and Boyd, and Edinburgh and Snedecor, *Statistical Methods*, Iowa University Press. Reproduced by permission of Prentice-Hall, Inc., Englewood Cliffs, New Jersey.

APPENDIX 10

Values of ρ for $\alpha = .05$ and $.01$ (rank-order correlation coefficient)

N	.05	.01
5	1.000	–
6	.886	1.000
7	.786	.929
8	.738	.881
9	.683	.833
10	.648	.794
12	.591	.777
14	.544	.714
16	.506	.665
18	.475	.625
20	.450	.591
22	.428	.562
24	.409	.537
26	.392	.515
28	.377	.496
30	.364	.478

Adapted from J. T. Spence et al., *Elementary Statistics*, 2nd ed., Appleton-Century-Crofts, 1968, p. 237; derived from E. G. Olds, Distribution of the sum of squares of rank differences for small numbers of individuals, *Annals of Mathematical Statistics*, 1938, 9, 133–148, and The 5% significance levels for sum of squares of rank differences and a correction, *Annals of Mathematical Statistics*, 1949, **20**, 117–118. Reproduced by permission of Prentice-Hall, Inc., Englewood Cliffs, New Jersey.

REFERENCES

Bresnahan, Jean L., and M. M. Shapiro. A general equation and technique for the exact partitioning of chi-square contingency tables. *Psychological Bulletin*, 1966, **66**, 252–262.

Campbell, Donald T., and Julian C. Stanley. *Experimental and Quasi-Experimental Designs for Research*. Chicago: Rand McNally, 1966.

Castellan, N. J., Jr. On the partitioning of contingency tables. *Psychological Bulletin*, 1965, **64**, 330–338.

Cicchetti, Dominic V. Extensions of multiple-range tests to interaction tables in the analysis of variance: A rapid approximate solution. *Psychological Bulletin*, 1972, **77**, 405–408.

Edgington, Eugene S. Statistical inference and nonrandom samples. *Psychological Bulletin*, 1966, **66**, 485–487.

Edwards, Allen J. *Experimental Design in Psychological Research*. Third edition. New York: Holt, Rinehart, and Winston, 1968.

Games, Paul A. Multiple comparisons of means. *American Educational Research Journal*, 1971, **8**, 531–565.

Harshbarger, Thad R. *Introductory Statistics: A Decision Map*. New York: Macmillan, 1971.

Hays, W. L. *Statistics for Psychologists*. New York: Holt, Rinehart, and Winston, 1963.

Kimmel, H. D. Three criteria for the use of one-tailed tests. *Psychological Bulletin*, 1957, **54**, 351–353.

Kirk, Roger E. *Experimental Design: Procedures for the Behavioral Sciences*. Monterey: Brooks/Cole, 1968.

Kirk, Roger E. (Ed.) *Statistical Issues: A Reader for the Behavioral Sciences*. Monterey: Brooks/Cole, 1972.

Ryan, T. A. Multiple comparisons in psychological research. *Psychological Bulletin*, 1959, **56**, 26–47. Reprinted in Kirk (1972).

Ryan, T. A. Significance tests for multiple comparison of proportions, variances and other statistics. *Psychological Bulletin*, 1960, **57**, 318–328.

Segal, S. *Nonparametric Methods for the Behavioral Sciences*. New York: McGraw-Hill, 1956.

Senders, Virginia L. *Measurement and Statistics*. New York: Oxford University Press, 1958.

Spence, J. T., B. J. Underwood, C. P. Duncan, and J. W. Cotton. *Elementary Statistics*. Second edition. New York: Appleton-Century, 1968.

Vaughn, Graham M., and Michael C. Corballis. Beyond tests of significance: Estimating strength of effects in selected ANOVA designs. *Psychological Bulletin*, 1969, 72, 204–213. Reprinted in Kirk (1972).

Wike, E. L. *A Statistical Primer for Psychology Students.* Chicago: Aldine-Atherton, 1971.

Winer, B. J. *Statistical Principles in Experimental Design.* New York: McGraw-Hill, 1962.

INDEX